YOUNG BAYARD

By Duane R. Ethington

Dedication

This book is dedicated to some of my very special students in both Tae Kwon Do and ISS-HOGAI JUJITSU through my 55 years of teaching.

Along with my son Sean, they are:

Paul Rosencrans, Randy Harper, Chris Fowler, Chris Houtchens, Scott Rose, Rosalio Zapata, Jeff Jackson, David Coleman, Jeff Bys, Matt DeFord, Tony Wright, Mike Garcia, Bobby Blakey, Milton Garcia, Chris Fowler, David Arehart, Mark Wiltfong, Jason and Candace Wigley, Desiree Day, Jerry D. Smith, Brice Buckley, Kenny Robbins, Carlo Falco, Yogi Ybarra, Joseph Ybarra, Audrey Escobedo and Angel Montano.

Prologue

The peaceful and most often tranquil valley of Graisivaudan had always been the pride of all Dauphine. A picture book setting, tucked away in a corner of the southeastern part of France, the valley extended from the towns of Chambrey to the land beyond Grenoble. The valley was generously irrigated by the swift running Isere River which threaded its way lazily through the length of the valley. Rich in fields of grain and fertile vineyards the valley was a spacious corridor nestled between bold mountain ranges that were beautifully capped with an abundance of glittering snow.

The fabled valley was rich in something else, as well. Pride and tradition reigned supreme here. The entire valley was strewn with the noble houses of the finest family names in all France. The houses of Theys, Comier, Boissieu, Acres, Montaynard and Guiffreys shared the limelight with the most distinguished houses of Alleman, Bocsozells, Beaumonts and the Terrails.

Each of these mighty lords were legendary throughout the far reaches of France as the staunch and loyal followers of the King. These lords of the valley, to the man, were men of whose chief pursuit of life was war. Their feared shields traditionally formed the very bulwark of French military might and could be seen, always, in the front ranks of any battle.

One of these great manor houses was a magnificent structure known throughout the land as Castle Bayard, home of the proud and esteemed Terrail family, whose history was steeped in honor and glory. The entire history of the Terrail's was proudly displayed on the walls inside Castle Bayard, itself. It was a history that began when the Terrails controlled only the land of Grignon; many years before they added the lands of Bermin and Bayard. It was a legendary and most envied history, which saw the head of the family, for almost three hundred years, fall in battle in the service of their king and country. It began with Aubert, who had fallen at the battle of Varay in the year 1325. Following Aubert was Robert, who met honorable death at the feet of his king at the battle of Les Marches in 1337.

After Robert came Phillip and his two sons, Pierre I and John. Phillip had been killed at Poitiers in 1356. His son, Pierre I had built Castle Bayard in 1403 and then had fallen courageously at the battle of Agincourt in 1415. Pierre's brother, John was killed at the Battle of Verneuil in 1424.

Then came one of the most renowned of all the Terrails; Pierre II, who was known as "The Sword" who fell in death at Montlhery with six mortal wounds in a great battle in 1465.

Pierre II's younger brother Jacques founded the famous cadet branch of the Terrails of Bernin and died in battle in 1470. The present Lord of Bayard, Aymon Terrail received an enemy lance in his chest that now left him permanently crippled with a limp and useless right arm. He had many nightmares about that fateful Battle of Guinegate, which of course, was his last. The Terrails of Bayard were related, by birth, to the powerful and distinguished families of Acres, Bocsozells and Theys and by marriage to the very respected house of Alleman de Lavell. Lady Helen Terrail Bayard, wife of Aymon was the sister of the Bishop of Grenoble and from the house of Alleman.

It is in this storied valley among these elite Frenchmen that we begin our story.

CHAPTER ONE

The harsh north wind blew in gusts as it entered Savoy from between the high, rugged slopes of snow. It whistled its chilly sonnet throughout the valley. The wind seemed to be particularly active this time of year as it whipped around the secluded manor house known as Castle Bayard.

Castle Bayard was a magnificent structure mounted on top of the long hill which offered a clear view of the entire valley below from any direction one desired to look.

Lord Aymon Terrail was now over seventy years of age, stooped and sporting his withered right arm. Twenty years earlier he had miraculously escaped the clutches of death when he received a near fatal chest wound from an enemy lance at the battle of Guinegate. Though he lived, the harrowing experience had left him permanently crippled with that limp and useless right arm, as well as an ugly scar on his left thigh from an enemy arrow. That battle ended a brilliant and esteemed military career that few in France could surpass at that time. No longer able to be active in the battle for his country, the old and honored soldier now found much time on his hands. It was time spent reflecting on glorious, long gone by, battle-filled days.

Days, too, to spend with his family, which included his wife, Lady Helen and his children, George, Pierre, Phillip, Jacques, Marie, Jeanne, Catherine and Claudia.

Nearly every evening, the old lord would sit by his huge fire place and tell stories to any or all that would stay within earshot of hearing. However, the old man particularly delighted in spinning his tales to his children. Lord Bayard was a master spellbinder, to be sure, and always found a willing and eager appreciative audience. Perhaps, though, Lord Bayard's most enchanted and ardent listener was his own thirteen year old son, Pierre. The lad with the dancing grey eyes would sit for hours and seemingly never tire of listening to the stories of his father's great yarns concerning various battles and fabled kinsmen. The boy especially enjoyed the tales of Charles of Burgundy and of Dubois and of Du Gresclin, Chandos, Talbot and of the soldier maid Joan.

As much as Pierre Terrail loved these stories, though, he reveled still more in the stories that concerned his own proud and glorious warrior family. More likely than not, the evening would almost always wind up with the old lord and Pierre and the exciting accounts of their brave ancestry. A glow of pride would always flush the heart of Pierre when he would hear of his own grandfather, Pierre II, who was known and revered as "The Sword". Lord Aymon had always told Pierre of how he so closely resembled his famous grandfather in both temperament and features.

Young Pierre knew each story by heart. The stories, undoubtedly, spawned the dreams and visions that fueled the desires of the youngster to follow in the footsteps of his famous ancestry.

It was exactly on one of those such evenings, just after the main meal and before the story session, that old Lord Aymon sat before his fire place. His stern blue-grey eyes looked thoughtfully into the flames which licked up the side of the stones. Surrounding him, the multicolored tiled floors were strewn with intricate patterns of images of magnificent horsemen, their great animals standing boldly out. The walls of the huge hall were heavily hung with a myriad of brilliantly colored banners, giant tapestries and standards which glistened brightly beside various suits of armor.

Along both sides of the great walls of the passageway hung shields and armor or great clusters of lances, swords, daggers and other weapons of war. Here, too, hung scattered oils in likeness of various great warrior members of the family Terrail.

Sitting silently in his high backed chair, the old lord stroked his full white beard. It was a habit that he engaged in whenever he was pondering a particularly difficult problem. The thoughts that now danced around in his head concerned the exciting days of his youth, which could never be brought back, mingled with the deep and growing concern of the future of his children. This subject was one he had dwelled on quite a bit of late. His cough was not getting any better and his old war wound seemed to grow more troublesome with each passing day. Lord Bayard had to face the fact that his days on earth were beginning to be numbered. Before he departed, his deep desire was to see that his beloved children were set on the correct footpaths for their lives. This especially was true with the thoughts of his four young sons.

With a sudden and startling warmth of anticipation in his voice, Aymon Terrail turned toward his wife and vigorously spoke out. "Bon Dieu!" he exclaimed in a voice that took Lady Helen and her two handmaidens quite by surprise, causing them to jump a bit with start. "Of a truth, I will send for my sons. The time has come for me to speak with each of them concerning a most important matter."

Lady Helen, looking up over her needlework, frowned playfully at her husband. "Now what is so important as to cause you to nearly frighten us poor women to death?"

The old lord smiled, realizing that his abruptness had been a shock to the nerves of the women. He leaned over and placed a soft kiss on her brow. "Forgive me, my love." he apologized quietly. "But I am excited. I would speak with my sons and learn of what way in their lives they might be thinking of letting their footsteps carry them."

Lord Bayard reached for the hanging stick near his high backed chair and struck the small gong a healthy blow. Instantly, two man servants entered and gave a slight bow before the lord. "My lord?" asked one man.

"I would that my sons have council with me." instructed the old soldier. "And I urge that they come quickly."

Lord Bayard needn't have added the last phrase for whenever he sent for them they all knew to come as quickly as possible without being told.

The servants dismissed themselves and quickly set out to complete their instructions.

With cheeks flushed, the four youngsters bolted through the huge door and entered the great hall. Panting from their haste, the four boys, each in turn and according to his age and custom, kissed their mother on the cheek and came to stand reverently before their father's high-backed chair.

The first long and agonizing moment passed in complete silence as Lord Bayard sat studying the expressions on the bright faces before him. Slowly, the old man's stern, wrinkled face broke into a huge smile, a gesture that seemed to cover the entirety of his face from his neck to his hair. The old soldier chuckled a special chuckle that was reserved only for his sons and waved his hand to indicate that they should sit.

"Come! Gather about me, my sons." offered the old man.

George, at sixteen and being the eldest, sat at the right hand of his father, the proper place for the eldest. Pierre and the others gathered around their father's foot stool and anxiously waited to hear what he was going to tell them this time.

Old Aymon looked, again, into the eager bright faces and proudly spoke. "You, my young warriors" he said. "The time has come in your lives for you to be giving serious thought to your futures."

Lord Bayard let the boys reflect on his words for a long moment, each in his own way. Then he turned to his oldest son. "George?" he prompted. "What is it you wish to do with your life?"

George, a quiet boy, thin of body and meek of

temperament, gulped and cleared his throat. He looked at his brothers and then at his father and answered.

"My father, it has always been my fondest dream to, someday, take charge of the order of our esteemed house, Castle Bayard."

"Yes. Well spoken." agreed the old man as he reached his left hand across his body rather awkwardly to pat his son on the shoulder. "You have long shown an aptitude for the complexities of being lord of the manor. We shall accelerate your training in this quest."

Satisfied with George's answer, the old soldier then turned a loving face toward young Pierre. Looking into the innocent, yet straight forward grey eyes, the man of the house asked "You, my young warrior? You who have the temperament of a lamb and the heart of a lion, have you given thought toward your future?"

Pierre's bright face suddenly faded into that of a very solemn and serious young man. As serious as a thirteen year old boy could muster, of course. The boy thought for a long moment. It was obvious that he was trying to capture just the right words with which to answer his father's very serious and important question. His grey eyes gleamed as he squared his broad shoulders, took a deep breath and gave his answer with unusual straight forwardness for one of his tender years.

"My lord and father." began young Pierre. "Long have I known what my heart desires. I have felt that my destiny doth lie in the estate of bearing arms as was chosen by you and my grandfather and his father before him. Although by your paternal love, I feel myself so greatly bound that I ought not be forgetful of all else and serve you till your life is at end, I must answer truthfully. There is room for only one lord of the manor, my father. Of a truth, I do not believe myself to be of the mold to linger around the manor. The stories concerning the great deeds of yourself and our ancestors have taken such root in my heart that, methinks, there is naught else would satisfy me."

Pierre paused to make sure his father was listening. Indeed, his speech had been so eloquent and unexpected that everyone in the room was riveted to his every word. Even the old lord tried to hide his slight smile of pride.

"Therefore, if it pleases you, my father" continued Pierre, "I wish only to join in the pursuit of arms. For it is the one thing in this world that I most desire. By God's help, I intend to bring you or your house no dishonor or shame."

Lord Bayard looked down at the youth with the square cropped brown hair and the old face brightened with immense pride. Slowly, a broad smile cut a swath across the wrinkled and bearded face and he fought the tears that were already trickling from the corner of his eyes.

"A noble and most fitting answer, my son," said the old soldier. "May God, indeed, give thee His grace. You barest such resemblance in both face and figure to your grandfather, who was in his own time one of the greatest and most accomplished knights in all Christendom."

The old lord smiled a proud smile and continued. "It would not surprise me, at all, if you performed such deeds that would cause you to outshine even him before you are through with it."

Pierre could feel his cheeks blush as a rousing agreement came from the others in the great hall. "I will, forthwith, endeavor to further your desires."

Pierre's young face lit up with an indescribable smile of happiness. His father's dreams and stories had, to be sure, influenced him to his decision. Yet, the immediate condition of the very times he lived in had also helped to guide the fate of the youngster from Castle Bayard as it did a great many boys of his age. France continually needed soldiers in any number of situations where enemies from both within and without threatened peace and their very lives at times.

Little did young Pierre Terrail of Bayard realize

as he stood there before his family and barely hearing the words of his two younger brothers who both swore to follow focused footsteps of other religious leaders of the family, that he was destined to become one of the most renowned, best loved and most accomplished soldiers of all time from any land on the face of the earth.

The very next day, old Lord Bayard sent a rider off to Grenoble to call on his brother-in-law, Gallia Christina, the Bishop of Grenoble.

Less than a week later, the Bishop arrived at Castle Bayard amid great fanfare and brought with him his constant entourage. Lord Bayard had also invited some other much loved kinsmen to help celebrate this very special occasion.

Young Pierre could hardly sleep that night as his father and the Bishop sat up most of the night talking and calling each other "important men of Dauphiny". Pierre knew, too, that no small portion of their conversation would be centered around him and his choice for his future.

The youth was up long before the sun had topped the hill and even before the Bishop held Mass. It was Pierre, himself, who served breakfast to his father, the Bishop and a dozen other personages who were family and special guests.

After breakfast, the old lord was almost bubbling

with joy and excitement as he addressed the elite gathering from his high backed chair.

"My lords and gentlemen." his voice boomed out after getting their collective attentions. "The reason I have summoned you all here is a very happy one for the House of Bayard. All present are my kinfolk and friends. You all, therefore, know the state of my health. I am so feeble and hampered by it that I should be surprised if it became well-nigh impossible for me to live two years longer."

Many guests moved restlessly in their seat or where they stood around the great fire place. None wished to think of such a possibility even though they, almost daily, considered such could come into the lives of any one of them.

The old lord did not want to dwell on that negative thought, so he continued.

"God has graciously granted me four fine sons. Of each one, I am most proud and pleased. I have inquired of each as to what their thoughts on their future does hold.

Amongst them, my son Pierre has disclosed to me his great desire to follow the noble and very fitting pursuit of arms as has so many of his ancestors before him."

A cheer and pounding on legs, stone walls or each other's back and shoulders followed the words. Every man in the room secretly wished for his sons to want to follow in the noble footsteps of their famed ancestry.

"It is needful of me" continued the old soldier when the noise subsided, "as you are all aware, in the beginning to place my son in the house hold of some great prince or other lord as myself, I am unable to instruct his proper education in the matter of arms. I, therefore, do entreat each of you to give me council in his behalf on this important matter. With whom I might best place him?"

After exchanging quick glances and mumbling words of encouragement all around with each man knowing Pierre would deserve a much finer schooling than any of them could give, one young gentleman from the House of Bocsozell spoke up.

"Methinks" said the man "that young Terrail will have to serve His Majesty himself, in order to gain proper and the best training of it."

Of course, there was a loud and immediate, to the man agreement from around the room. Yet, each man was well aware of the fact that Pierre could not simply embark on Page hood by beginning in the King's own palace unless he was of the king's own Royal family.

Thus, several suggestions were put before Lord Bayard as to which great lord would best serve to tutor his young warrior to be. Lord Alleman, the older brother of Lady Helen and also a Bishop, stood and cleared his throat as all eyes turned toward him.

"My kinsmen." began the Bishop with a deep voice that boomed through the great hall. "I submit that I have, indeed, come upon the answer to the problem before us. I think the great House of Bourbon should be a right fine place for the lad to gain his tutoring. Everyone knows my lord Bourbon has an excellent school for any beginner at arms." Almost to the man, each person in the great hall silently exchanged knowing glances of approval and quick nods of agreement.

"My lord, Bocsozell has made a most excellent choice" agreed Lord Bayard.

After a small round of applause, another distinguished gentleman in the room, stood. It was old Lord Theys. "Yes, the House of Bourbon and the House of d'Ligny are both excellent and are among the finest schools of arms in all of France. Each is a school that any boy would be privileged to attend. After due thought on this matter, however, and considering the importance to Lord Bayard of his son carrying on the proud name, methinks there is even one school which would be better."

"Better?" came several mumbles all at once.

"Yes." assured Lord Theys. "Methinks that the young lad will be best served in the house of our friend, Charles, the young Duke of Savoy." After a brief period of collective silence as each man mulled over the ramifications of the suggestion, sudden and exuberant nods and murmurs of total agreement prevailed.

"What a marvelous suggestion" agreed Bocsozell and Lord Alleman together.

"Are there any words of opposition?" asked old Lord Bayard as he gave a quick glance around the room. None in the room could think of anything to say opposing the suggestion. The dashing young Duke was quickly gaining respect among his peers for the turning out of prized fighting men.

Lord Bayard slowly nodded his approval to no one in particular. "It is done, then" he said with a certain degree of finality.

Thus, in that moment of agreement, young Pierre Terrail of Bayard's future was determined. From that point forward the process would simply be for Lord Bayard to follow formality and send a letter of request to the young Duke of Savoy.

"Yes, the matter seems to be sufficiently settled" agreed another lord of the valley, Lord Vriage, who was the son of the Lord Soffrey.

"The youth cannot go to the House of Savoy without a proper mount" spoke up the Bishop of Grenoble. "I have an excellent beast that will serve a young pupil right well, methinks."

"Good!" lamented a happy Lord Bayard. "I will

send a letter to Duke Charles, forthwith, and when he answers we will all gather in this place again and see to it that the young man is given a proper send off."

It was a very proud and exciting day in the House of Bayard when the old soldier received a reply to his inquiry of the Duke. With anxious fingers, the old man tore open the seal on the letter. His excitement was almost uncontainable as he began to read the Duke of Savoy's answer to Lady Helen.

"My Good Lord." he read. "It is with extreme pleasure and deep honor that I tell you to send your son, Pierre IV, as a ward of my humble house for the purpose of furthering his education in arms toward the preparation of his future knighthood. In this year of 1486, when all of Europe stands at much unrest, it is a proper time to think of the future of France through the molding of her youth. A place is awaiting young Lord Bayard whenever you choose to send him. Cordially, Charles, Duke of Savoy."

The old soldier sprang from his chair. "Good wife," he shouted excitedly, "Pierre has been accepted at the Court of Savoy. He shall leave on the morrow for his new home. I will send for him and tell him of the good news."

Tears of mixed emotion trickled down Lady Helen's face as she thought about what was taking place.

Old Lord Aymon saw her mild distress and moved over to her. He gently put his arm around her shoulders. "Now, now, my dear," he said in as gentle a voice as he could muster, "don't let the children see your tears. Pierre is a big boy now. It is time for him to receive the proper training which he will need very badly if he is to survive the future of our country against its enemies. We must accept this as a grand opportunity for the betterment of his future. We must be strong and face up to losing him. It must be done for the boy's own sake."

Young Pierre could scarcely contain his joy when he received the news about the letter from Savoy. The entire family was proud and happy and with good justification. Pierre was already an accomplished horseman far above his years and he seemed to have a real talent when it came to addressing any weapon that his father would allow him to touch. Now there would be another Bayard to carry the exalted name into battle for God, country and king once more.

The remainder of the busy day was spent in preparing the youth for his long journey to Duke Charles's present Court at La Perouse.

When the evening meal was at hand, it proved to

be a festive occasion, indeed. The long wooden table was piled high with rich and exciting dishes. There was bear steak, venison, the shoulder of a wild boar, roast swan, chicken and pigeon pie. The kitchen had certainly gone all out for the young traveler. Lady Bayard, personally, supervised the entire affair and saw to it that her son had all his favorite dishes, which included eel pie.

After this elaborate setting was heartily and thoroughly devoured by the family and guests, they were then treated to a fancy array of sweat meats, cloved ginger and other spices. Pierre requested another dish of frumenty, a tasty dish of milk, sugar and wheat. Then, following the musicians came the ballad singers and story tellers. The evening grew long and ended with the court jester making his entrance and enticing the crowd with jokes and buffoonery.
    Finally, everyone grew very sleepy and craved to be tucked deeply beneath some warm covers for a good night's sleep.

    Pierre could not remember when there had been a more enjoyable evening in the big house. Perhaps, he reasoned, it was because he had never been so personally involved before. He did not remember, though, one of his father's strong young knights carrying him up the long flight of steps to his room high in the circular tower.

The following morning seemed to be a continuation of the night before with everyone excited and keeping very busy coming by to tell Pierre goodbye and wishing him safe travel.

Pierre's uncle, the Bishop of Grenoble, kept his word and proudly led a beautiful black horse up to Pierre.

"A present for you which I got from the good Lord of Vriage" said the Bishop. "He will serve you well. Try him out."

Pierre could tell by the look on everyones' faces that they expected him to be afraid of the big horse. But he wasn't afraid. He was a good horseman and even could outride some of the knights, who were all good horsemen. Pierre climbed aboard without hesitation and spurred the animal. Then he delighted everyone by riding in great circles around the group as they applauded.

"He truly has the experience of a man of thirty" called out the proud Bishop.

Pierre dismounted, a very proud and happy boy.

"My lord, uncle," he said, "I hope, that with God's help, before six years are over, to be able to guide this or any other horse in more dangerous places. For here I am, in the midst of friends and I may, then, be among enemies of the master I shall be serving. Thank you for this fine animal which I now dub 'Wingfoot' because of his great speed."

All applauded the selection of the name as Pierre turned toward his father.

"My lord and father." he said. "I pray of God that He may grant to you a happy and long life and to me the happiness that before leaving this world, you may have good tidings of me."

Pierre hugged his father and then turned to his mother, who had been standing quietly by, fighting her tears.

"Walk with me, Pierre" she said quietly.

They stepped aside from the others and began to walk very slowly.

"Pierre, my child," said the good lady as they walked, "you are going into the service of a noble prince. In as much as a mother can command her child, I command three things from you in your going. First: above all things, you shall love and fear God's service. Neither offend Him, if it be possible; for it is He that created us all, it is He that makes us live and it is He that saves us. Without Him and His grace we are unable to do one good thing in this world. Each night and each morning recommend thyself to Him and He will assist thee. Do you understand what I am saying?"

"Yes, my mother" assured the boy with a small

nod.

"Good! The second thing I command of thee is that you be generous and courteous to all men, keeping thyself from all pride. Be humble and useful. Be neither a liar nor a slanderer. Be sober in both eating and drinking and flee from envy, as it is an ugly vice. Be neither a flatterer nor a tale-bearer, for such men do not attain a great perfection in anyone's eyes. Be loyal in both word and deed. Be a man of your word, always. Help poor widows and orphans and God will surely reward you for it."

Lady Helen turned her young son's face to her and looked deeply into his bright eyes. "Thirdly," she continued, "with the goods that God gives you, be charitable to the poor and needy; for giving in God's honor impoverishes no man. Believe me, my son, that such charities will profit you much more in body and in soul than you can imagine.

Therein is all that I charge you with. I must think that your father and I will not live a long time further. May God give us, at least, the happiness whilst we still live of always hearing good tidings of you."

Pierre hugged his mother and then took her by her hand.

"Madam, my dear and beloved mother" he said quietly. "For your good precepts and faith in me, with all humbleness, I do thank you. I hope with all my being to be able to follow them so as to content you by the favor of Him unto whose keeping you do commit me. As for the rest, my mother, after most humbly commending myself to your good grace, I take my leave of you."

Lady Bayard reached out and gently stroked the soft brown curls of his hair while fighting hard to drive back her tears. The handsome young man hugged her and kissed her and then spoke once more. "Mother, no need to worry yourself. I shall make you and my father very proud of me."

"We are already very proud of you, my son" she answered as she pulled his head to her bosom and held him, perhaps, for the last time. Then the good woman, as all the women of her time were expected to do in a situation such as this, reluctantly but firmly pushed him away. She motioned for a man servant, who had been patiently standing nearby, to come to her. He brought her a small box. From the box she took a small silk scarf and tucked it inside Pierre's tunic.

"In here are six crowns of gold and two in change."

"But, my mother?"

"And in this box," she continued as she ignored his small protest, "are some linen which you will need. Also in here is a special gift for the Duke's equerry, that he might look after you a little closer in your first days at Savoy."

"But, mother, I need no special favors" said Pierre.

"Perhaps not" she agreed. "But take them just the same."

Both of their attentions were drawn to the sound of the horn's blaring blast as it echoed through the courtyard of Castle Bayard. It was the signal announcing that it was time for the young traveler to be on his way.

Pierre and his mother gave silent, knowing nods to each other and then he turned and was off.

A great surge of excitement rushed throughout his young body as every fiber of his proud heart pumped at such a rate as to make him feel as if the sound might be heard beyond the walls of the castle itself. Pierre hurried to stand before his father, once again.

Then, in turn, he bid his brothers and sisters a proper goodbye. Then came his favorite uncle, Laurent Alleman and two cousins. With this completed, Pierre stood before his parents once more.

"My father." he managed as he fought the lump he could feel rising somewhere in his throat. "I pray our Lord that he give you good life and long, and to me the grace that before death you may hear but good news of me."

The old soldier grinned broadly and held out his arm as Pierre flung his small frame against his father's chest and stomach. They hugged for a long moment. A slight crack appeared in the usually strong voice of Lord Bayard as he spoke.

"Go, my son. Know that we all send our love with you. Do yourself and the name of Bayard justice, as I know that you must. You are our champion, now, young warrior. Mount your steed and God speed you to glory."

With his father's words echoing in his ears, Pierre Terrail Bayard mounted his new horse, Wingfoot, waved goodbye to all in the courtyard and rode promptly and proudly from Castle Bayard.

Reaching the drawbridge, Pierre pulled his horse to a halt. He turned in his saddle and bravely gave one last look at the only home he had ever known. He gave a final wave to the people who, in turn, let out a resounding cheer. Pierre then cantered off, his six escorts riding two abreast behind him. Pierre was entering a new way of life that would inflame and, indeed, consume him till the day he would die.

CHAPTER TWO

They had been gone for four full days from Castle Bayard. Even Pierre, who loved to be in his saddle, was becoming bone weary. On the morning of the fifth day, the small group of riders came to a halt at the edge of a vast fertile meadow.

"Your destination, young sire." pointed out Sir Grainger, the knight in charge. "There stands the present Court of Savoy."

Directing his gaze across plush greenery of the meadow, Pierre saw that the man-at-arms was pointing to a cluster of structures just beyond a formidable looking wall.

Feeling the excitement mount within him, the lad's hand went to his tunic where his fingers probed with reassuring certainty as he touched the letter of introduction.

"We will leave you now, Master Bayard." added Sir Grainger. "It will be much more impressive if you enter your new home without us. You are a worthy man to enter yon gates and we will relate to Monsieur Bayard that you pushed us, even at night, to gain your objective." The knight from Castle

Bayard smiled at Pierre.

"Thank you, Mon Ami. I shall miss all of you," said Pierre in his most convincing voice with hopes that the knights did not detect the hint of insecurity that he thought he might be projecting. If they did, they said nothing.

"God speed you on your way back home" added Pierre.

"Stay well young Sir. We will meet again." said Sir Grainger with a nod and a warm smile just before turning his horse and moving back to join his waiting comrades. The riders from Castle Bayard turned their horses and moved off and out of sight in the thicket as they headed home, their mission completed.

Pierre took a deep breath and, once again, turned his full attention to what lay ahead. He brushed some of the traveling dust from his clothing and pointed his horse toward the opposite end of the meadow.

As Pierre drew within four hundred yards of the Castle, he heard the loud blare of a trumpet, giving warning of his approach. Slowly and steadily, the boy guided Wingfoot forward as he pulled the pack mule closely behind him. Quickly, almost too quickly for his liking, he closed the distance between himself and the towering walls before him.

Even at a distant glance, it was apparent that this

place was much larger than his home of Castle Bayard. The boy wondered just how he was going to fit into such a grand place as Castle Savoy.

Suddenly, a great clatter and clinking of chains drew Pierre's full attention toward the walls. The large drawbridge was being lowered.

Pierre could feel his heart pounding and his blood racing. He gripped the reins tightly in his sweating hands and gulped as all sorts of thoughts raced within his head. What if they attacked? What if this would be the wrong place and they would not let him enter? What if?

Slowly, the two riders eased their horses forward to close the distance between themselves and their visitor. When they were on either side of him the two riders stopped. They sat for a long moment and simply looked at the visitor in silence.

"Tis but a lad" exclaimed one of the riders in surprise.

The two knights then pulled their helmets off and pointed their lances skyward so as not to frighten Pierre any more than they probably already had done.

"What is your business at Savoy, lad?" asked the second man.

"I am Pierre Terrail of Castle Bayard, sir" answered the boy, not sure if his words were even getting through the cotton that seemed to be filling his mouth. "I have come to serve in the house of the good Duke of Savoy as Page."

The two knights looked quizzically at each other. They, of course, were familiar with the name of Bayard and knew of the steeped glorious history attached to it. They looked beyond Pierre toward the woods where Sir Grainger and the others had been before heading home.

"Welcome to Savoy, young Bayard, of course" said one man. "Where is your escort?"

"They left me at the edge of yon trees" answered Pierre as he felt himself growing more at ease.

"I am Darwin, knight in the service of the Duke of Savoy" said the second man who spoke. "Allow me to introduce my comrade in arms, Sir Phillip. You are in luck, young Bayard, for the Duke and Duchess, only yesterday, have arrived from their three weeks at Geneva. Come! Follow us."

The escorts led the way and the impressed boy from Bayard followed as instructed, little realizing that he was taking the first steps to an incredible military career of his own.

The trio of riders passed over the drawbridge, through the giant gates and into the huge courtyard, which was already filled with curious onlookers. It was always a big event when a visitor arrived at court. When the visitor turned out to be a small boy traveling all alone, the interest of everyone in the castle was piqued.

Pierre rode in his assigned place, between the two knights, as the tiny procession made its way to the great house.

A group of people stood on the steps of the place curiously observing the youthful newcomer. They stopped before the group and the two escorts nodded for Pierre to dismount. The lad gracefully did so.

Fighting the new lump in his throat, Pierre withdrew the parchment from its resting place beneath his tunic. Suddenly, he felt himself fully flushed of face as he realized that he did not know which of the people before him would be the Duke.

It was Darwin who saved the day as he motioned toward the dashing young man who was not much older than Pierre's brother, George.

Four firm steps carried Pierre to within two feet of the youthful Duke. Then, falling on one knee to keep with custom, Pierre offered his greeting.

"Your Grace," he said "a letter from my father, the Lord Aymon Terrail of Bayard."

Pierre watched intently as the Duke took the paper from him, opened it and then studied it for a moment. Charles was a rather handsome young man. He sported a thin, black moustache under a nose that was curved on a face that was slightly pockmarked, observed Pierre in silence.

When he had finished the letter, Duke Charles gazed down upon the still kneeling youth before him and bid him arise with a wave of his hand.

"Your father has sired a fine specimen of a son" said the duke with a small smile. "I am sure that he is well pleased with you and very proud of you as I know we shall become here at Savoy. You will learn your lessons well and you will honor the court of Savoy before it is over with. Your father and your family have long served France with honor and dignity. We shall expect naught else from you."

"Thank you, Your Grace" said Pierre, not knowing what else to say. "I will not disappoint my father or you and your house."

"In that, I am sure you speak the truth, young Bayard" affirmed the duke. "Now, may I present Lady Blanche of Monferrato, the new Duchess of Savoy?"

Pierre turned to the beauty beside the Duke, who could not have been over sixteen years old herself. Again, Pierre dropped to one knee, and being careful to choose just the right words, spoke again.

"Your Grace," he said, "the honor is unquestionably mine. Of a truth, it is rare to meet such beauty and grace face to face."

It was plain to see that the young duchess was touched by his words. She offered her hand and

Pierre took it gently in his and kissed the white silken glove, ever careful not to squeeze the delicate fingers too tightly.

"How utterly charming" noted the Duchess with a slight giggle. "I beg you arise, young sir." Then Lady Blanche turned to her husband. "Such manners are commendable, my love."

"Indeed." said the duke as he turned to the crowd and called out so that all could hear. "We are honored, this day, to have the son of Lord Aymon Terrail of Bayard, who is a most notable knight of France, to join our household as Page. This is Pierre Terrail Bayard and methinks he will excel in all things."

The people cheered and clapped and Pierre felt a strange warmth swell within him. These people were really making him feel welcome to his new home. For the first time, the newcomer became aware of some of the faces in the crowd. A broad smile crossed his lips as he spotted the familiar face of a distant cousin, Arthaud de Bocsozell, from his father's side of the family. He had seen the man only a year ago when a company of travelers passed through Castle Bayard. He recalled how his cousin had been on the way somewhere to serve as Marshall of the Logis. Savoy must have been that place, he reasoned.

Duke Charles turned Pierre over to his cousin then he and the duchess excused themselves and went to attend to further household matters.

After Sir Darwin assigned a man to take Wingfoot to the stables and assured Pierre that his belongings would be collected from the mule and brought to his quarters, Pierre walked with Arthaud. As they walked, Pierre learned that he had three more kinsmen dwelling in Savoy. There were two of his mother's cousins, Antoine and Hughes de la Forest. Antoine was now Chamberlain at court and also served as the governor of Nice. Hughes was employed as Major-domo of the household of Savoy and as head steward, he was kept very busy. Hughes had a young son named Guillaume, who was already a Page. Along their walk, too, Pierre was introduced to the court jester, a rotund little man called La Plaisant.

Arthaud turned Pierre over to Guillaume, who led him to a huge door down a long corridor.

"This is to be your quarters." explained Guillaume. "There are six other Pages besides us who share the room. Enter and meet the others. I have other chores to attend to."

Before Pierre could even thank Guillaume, he was gone. Pierre turned the huge handle and pushed the heavy wooden door open about half way, took a deep breath and entered.

The room was huge with two tables in the center and a row of beds lining each wall. Two other young pages were in the room. A tall, muscular boy of seventeen, who stood a good three inches taller than Pierre, stood and walked over to him. "You must be the new one" he said. "I am Henri. I am in charge of the Pages in this place."

"I am Pierre Terrail of Bayard" said the newcomer with a smile.

"Your bed will be there, in the corner" said Henri as he appeared all business while pointing to the designated wooden bed with two thick mats of straw on it. "Come! The first thing we must do is get you properly clothed."

Pierre followed Henri to the fitting room where he received all the proper clothing he would be wearing as Page of Savoy. His uniform was of the red and grey coloring of the House of Savoy. He received an outer grey tunic, which could have been worn with or without a hood. Pierre thought he was all set but soon discovered that there was much more coming his way. He was also given a doublet and crimson hose and a red cap. This was just for everyday wear, he was told. Too, he was issued a special violet colored doublet of satin for special occasions. With that, he also received a satin lined tunic of black velvet with wide sleeves of crimson and some silver hose.

"In the case of you being summoned to mourning." said Henri. "You will attire yourself in all black" The head Page handed Pierre the final items, two pairs of black, double soled shoes. Then Henri accompanied the new Page back to his quarters. The head Page never stopped talking as he issued order upon order and added advice of his own. Pierre wondered if he would ever be able to remember everything.

Pierre put his double armload of clothing on the bed. When Henri headed for the door, Pierre sighed a small sigh of relief. At the door, the head Page turned. "I must get to my chores. Amuse yourself till the noon meal."

"I will accompany you" offered Pierre.

"No need for that" assured Henri. "Today you had better take to your rest and enjoy it. On the morrow you will begin your toil soon enough." With that, Henri was gone, closing the door behind him and leaving the newcomer completely to himself.

Slowly, Pierre eased his weary body down on the bed. After several nights of sleeping on the ground in the forest, a bed was a welcome sight. Before he realized it, the boy from Bayard closed his grey eyes and was fast asleep.

The newcomer's slumber was, too soon,

interrupted by the hand tugging roughly at his shoulder. Pierre opened his eyes to see Henri standing over his bed. "Come! It is meal time."

After the meal, Pierre wandered out into the courtyard after being assured that he was to enjoy the rest of the day. It was already hot, and with no clouds above, the weather seemed even hotter. Pierre wiped the moisture from his brow with his sleeve and squinted as he tried to shield his eyes from the bright rays. He breathed deeply; filling his young nostrils with the scents of the summer surroundings of his new home.

Searching the courtyard, Pierre's gaze fell upon the open doors of the stables which were situated at the far end of the courtyard from him.

Pierre's sturdy quick strides carried him toward the stables as he took in all the sights about him. As all Bayard's did, Pierre loved animals and prided himself in taking good care of his horses. This horse was particularly dear to him as it was a gift from his favorite uncle. One sweeping gaze found the horse in a pen with two other horses and leisurely grazing from the hay in the huge trough.

"Here you are" called out Pierre as he walked up to Wingfoot after squeezing through the wooden railings of the pen. His hand gently stoked the horse's white mane. "How are you doing,

Wingfoot?" he said softly. "Are you getting enough to eat?"

"You have a very fine animal, there."

The unexpected sound of the voice had come from the doorway of the stables and startled the young newcomer from Bayard for a moment. Pierre whirled in the direction of the sound and discovered himself facing a tall young man in his early thirties. The broad shouldered man with the beard and suntanned face stood at the railing and smiled down at the newcomer.

"Thank you, sir" answered Pierre.

"I am Jacque." introduced the man as he stuck out a big hand. "I am Duke Charles' black smith and personal horse trainer."

"I am Pierre Bayard" responded the boy as he took note of how small his hand was inside the smith's own.

"Yes, I saw you arrive this morning" said the man. "A new page? You will do fine."

"Do you really think so?" asked Pierre as he looked for some kind of reassurance to offset all that Henri had told him would be expected of him.

"Of a certainty" answered the Smith. "Not only that, but I wager you will grow to be a fine knight one of these days, as well."

"Good Jacque." said Pierre. "I truly hope you are right. For it is my fondest wish for it to be so and to serve my God, my country and my king."

"Well spoken, young Bayard." said the smith as he reached over and ruffled the boy's hair. "Our country will surely be safe with men like you riding under her colors."

Feeling just a little embarrassed, the way he had always done when someone was paying too much attention to him, Pierre quickly changed the subject. "Do you think it would be permitted for me to take my horse for a ride?"

"I see no reason for anyone to be against it." assured the black smith. "Here, let me help you saddle him."

"If you do not mind, sir smith" countered Pierre, "I shall saddle him, myself. I have always done so. If a man cannot care for his own horse, then, pray what kind of a man is he?"

"Well... yes" answered Jacque with a half hidden smile. "This is quite a grown up philosophy for one so young. But, surely, you must see that it would not be purely right for a knight of France to be reduced to the position of stable boy? Besides, every knight has a good friend who would lend him a helping hand from time to time. So, with your permission, young Bayard, I will be your helping hand today."

"You drive a sound bargain of it, sir, Smith." conceded Pierre. "But, I am not yet a knight."

Jacque rubbed his hand through Pierre's head

again and thought of how much he already liked this boy from Bayard who had such a smiling face and such a refreshing personality. "Let us suppose you are in training and see what happens."

Pierre sat comfortably upon Wingfoot as the animal moved with flawless motion from the stable. The newcomer waved courteously to various folks as he passed them. Pierre had always felt at home when he was on horseback. All eyes seemed to turn toward this handsome young stranger. Everyone loved to see a good rider and his horse work. Curiosity drew them to this young newcomer.

Pierre leaned close to Wingfoot and spoke softly in the animal's ear as he eyed the corral next to the east end of the courtyard which stood three rails high. "Okay, big boy! Let us take yon fence. Show them your best leap."

As if Wingfoot understood his master's every word, the great horse sped up his gate, and at Pierre's signal, went over the high corral fence with considerable ease. Some of the onlookers applauded vigorously. Almost without hesitation, Pierre leaped his horse back out of the corral and into the courtyard again.

Then the boy from Bayard proceeded to treat the people of Savoy to an exhibition of riding, the likes of which they had not seen in a long time. Soon, a large crowd had gathered to watch the remarkable prowess of this fine youthful rider. Pierre bobbed and weaved in the saddle, hung from one stirrup while retrieving something from the ground, leaped from the saddle on a dead run and then bounced from the ground back into the saddle without Wingfoot even slowing down. The boy hung upside down from his saddle and crossed beneath the galloping horse and stood erect upon the saddle.

There were none in the courtyard who could boast of ever having witnessed better riding anywhere. This was truly a magnificently gifted boy, they would say to each other. Already he showed that he could ride better than many full grown knights.

As the people in the courtyard stood applauding, Pierre reined in Wingfoot at the stable door and dismounted without the use of his stirrups. Jacque was there to meet him and applauded along with the others.

"Excellent" approved the smith. "Somehow I just knew that you were no ordinary lad on a horse. Where did you learn such skill, young Bayard?"

"Of a truth, my father began my riding education when I was but five years old" answered Pierre. "But I must confess that I owe much of my skill to a gallant knight who visited my home at Bayard, when I was only nine. He was clad all in black and he showed me some of the more difficult things that

I was just doing."

"Who was this teacher?" questioned the smith.

"I know not, friend Jacque" answered Pierre. "But they said he was an English nobleman of great station and requested to be unknown. More I know not"

"Pierre Bayard?"

The voice was unfamiliar to Pierre but quite friendly. Jacque and Pierre both turned toward the author of the voice. Pierre was very surprised to see young Duke Charles, himself, walking toward him. The boy instantly bowed low.

"Your Grace?" said Pierre. "Did I do something wrong?"

"Wrong? On the contrary, young Bayard." assured the Duke. "It was most pleasurable for me to witness such outstanding riding. It is some of the finest I have ever seen. I sincerely mean that, young Bayard. You have a remarkable talent. Someday, perhaps, you will teach me?"

"Of a truth, it would be an honor, Your Grace." replied Pierre. "However, I must confess I was showing off a bit."

"If I had that talent, I would show off, too" said the duke with a smile. "I am taking most of my court on a long journey soon. I wish you to accompany me, young Bayard."

Without another word, Duke Charles turned and

was gone as swiftly as he had arrived.

"Did you hear that, my young friend?" asked Jacque with no small excitement in his words.

"That is great! You are one lucky young man. This not only is a great honor that the duke extends to you but it is also a fine opportunity. You will find that this is a very active court, indeed, as it is filled with youthful ambitions. The duke is young and has a young wife. They stay busy and you will do well with them. Methinks, good Page, that you are destined to live a charmed life of it."

Pierre heard the words but his youthful inexperience kept them from really taking root. His mind was on other things. He cared more about new friends than the monumental opportunities that traveling with the Duke of Savoy could offer. At least, for now, that was true.

"You, good Jacque" said Pierre, "will you be going on this journey?"

"Of a truth, yes I will" answered the smith. "I must look after their mounts. That is my job." Jacque reached for Wingfoot's reins. "I will curry him, personally."

"Oh, no" protested Pierre. "That is my job."

"Nonsense!" replied the black smith. "At Savoy, it is my job. That is what the duke pays me wages to do. Besides, it would not look right for a future knight of France to be playing stable boy. Now, no arguing! Off with you. Your steed will receive the finest care possible."

"Thank you, friend Jacque" said Pierre as he was thinking of what a very different life this would be from the one he was use to at Castle Bayard.

Pierre Bayard had been fairly warned that his training would be tough.

Tough it was. There were days when he could barely drag his weary body into his bed. As physically demanding as his training was, it was also a very exciting atmosphere to be in. He spent most of his time in the world of physical exercise and sports, of hunting, hawking, jousting, wrestling and swordsmanship. This activity all molded together and began to provide the young man from Bayard with the physique that would resist fever, weariness and wounds which would preserve him during forty years of battles to come. It was the groundwork which would temper his very soul in becoming one of the greatest of all fighting men France had ever or would ever know.

Pierre had a full schedule, indeed, in learning to master 'the woods' and learning to become an excellent hunter and hawker as well as learning the art of concealment and tracking on top of all his weaponry.

Riding was most important for any knight. A knight who could not sit his horse properly in battle would have no more battles. It soon became apparent that the youth from Castle Bayard could be rivaled by absolutely no one when it came to skills with a horse. He was an exceptional student. In most cases, he was already better than his three riding instructors.

Not only was Pierre, himself, an unvanquished rider, but his horse, Wingfoot was also one of the swiftest within the dominion of Savoy. Even the Duke's own fine knights of some reputation could seldom boast of a win in a race against Pierre.

Light weapons were introduced to the students. Months of diligent and rigorous training demanded from the Pages and Squires the use of the sword, lance, battle axe, bow, mace and dagger. The candidates also had to learn to become excellent swimmers. They were taught to box, wrestle and fence. Pierre especially enjoyed the times that the training permitted the holding of mock battles and contests of skills. This was often, as each trainee was expected to improve on all parts of his training. Each student gladly participated in every contest they could get themselves into as they eagerly wanted to please their instructors and their lord, the Duke.

These contests were always regarded with eager anticipation as everyone was always on the alert to which son of which famous knight would maintain and uphold the reputation of his valiant father. In this venue it did not take long for the young man from Bayard to establish himself as a favorite in most of the contests he would choose to enter.

Pierre's chief rivals became Henri de la Palu de Varembon, the chief Page who now was a Squire, Guillaume de la Forest, his own cousin and a tough youth named Lucquin Le Groing. More times than not, however, it would be Pierre Bayard who would come out ahead, even if it were ever so slightly, in the contests of skills.

Outdoor learning was not all that was required of Pierre. It was impressed upon him, over and over again, that he must be 'serviceable' to the ladies and lords of the court. He learned always to be gentle and polite and courteous. He learned all the proper things to say and the proper ways of doing what was expected of him.

Most of his indoor lessons were taught to him by the ladies of the court. These lessons included various dances, music appreciation, playing the harp, playing backgammon and greatly improving his chess. Although great stress was not put upon learning to read and write, Pierre learned even these lessons better than most. Soon too, Pierre Bayard swiftly became the favorite student among the ladies. His broad shoulders and slim waist and muscular body made him look older than his years.

He was, indeed, growing into a fine young man. The time seemed to pass exceedingly quickly. On top of all the new lessons to be learned were the constant journeys of the young Duke and Duchess.

Pierre's lifelong travels began with his first journey only three weeks after he had joined the House of Savoy. That long journey was made in easy stages over the mountains of Cenis and on to the rapidly growing city of Turin.

Life was proving to be an adventure, to be sure, for the lad who sent nights in the best of inns. These were famous inns of the time such as Ecude France at Aussois and the Sheep at Lanslebourg. Then on to the very famous Three Kings at Susa. Pierre also spent nights at palaces of Bishops and in fine houses of various gentlemen. The boy was bedazzled by the splendid receptions given at Turin for the Count of Duois, the son of the famous soldier of the same name and for Clara Gonzaga, whose son would later become the famous and powerful Constable of Bourbon. They spent the entire summer at Turin before venturing on to Moncrivello and then to Vercelli.

In the autumn, war broke out between the Duke of Savoy and the hotheaded Marquis of Saluzzo which prompted a quick trip back home.

The pages, like the other members of the household, were issued special winter clothing and were given very special duties to perform. Thus, only eight months away from Castle Bayard, young Pierre had begun what was to be one of the most glorious military careers in the history of mankind.

This first war proved to be an exciting and meaningful time to Pierre. Each side sported around a thousand combatants. The Pages were used as non-combatants, riding with messages and serving as liaison agents. It was in this way that Pierre's early training at Savoy was filled with violence, love of open fields, lusty comradeship and always accompanied by the uncertainty of unclear tomorrows. This campaign lasted well into April and ended just before Pierre's birthday, which was the twentieth. Even with all the unrest and constant confrontations with enemies, all the action did not come from outside the Duchy.
    Some of the trouble, in fact, sprang up from within the Duke's own domain.
    Pierre was awakened with a start at the blast of the trumpet from the south watch tower. Quickly dressing, Pierre hurried down the long flight of stairs and burst from the door into the huge courtyard. There he saw that the entire castle had been put on full alert with people scurrying around like members of a busy ant colony.
    Duke Charles, who had already been put into his

shield of protection by many swift hands, was moving across the courtyard toward the drawbridge. Sir Darwin was reporting to the duke just as Pierre came within earshot.

"They are demanding satisfaction, Your Grace." explained Darwin.

"Those knaves dare accuse the House of Savoy of stealing horses" spat the young Duke, who did not like his morning rest to be thus interrupted. "Are they alone?"

"Yes, your Grace." answered the man-at-arms. "It seems they are the only malcontents who seem to be in a sporting mood this fine morning."

"Then let us settle this so that I might enjoy my breakfast" said the duke, sharply. "Tell them that we will answer their challenge in equal number, by the pond a quarter hour hence."

Darwin quickly and unquestioningly set off to carry out his lord's orders. Duke Charles turned and saw Pierre and some others standing nearby. "Gather your horse, young Page." ordered the Duke. "I will have no other serve me in this affair."

Pierre gulped hard, blinking and then quickly gathered himself. He then excused himself and dashed toward the stables. All the time he ran, he was wondering if he would remember all of his lessons in what to do in this type of situation. He only understood that he was to assist the Duke in battle in any and every way possible. That was his duty.

"What if I am attacked?" he questioned himself aloud as he ran through the stable doors.

"What was that, young Bayard?" asked Jacque, who was in the first stall already saddling the Duke's battle charger.

"Nothing, friend Smith." answered Pierre as he remembered his teachers stressing that a page would be in no direct danger from an attack by a knight. For a man-at-arms who would attack a defenseless page would be shamed, disgraced and probably banished from his ranks by his peers.

Moments later, following Jacque, Pierre led Wingfoot from the stable. Already gathered by the draw bridge were the Duke's finest knights; Darwin, Phillip, Jon, Maurice and three others. Each man-at-arms had his page or squire riding behind him. Accompanying Pierre, who had the greatest honor of being 'Squire of the Body' to the duke, was Henri, Guillaume, Gaspard de Coligny, Lucquin Le Groing and three others.

As the draw bridge was lowered at the signal of the trumpet, Pierre wondered if the others had their hearts seemingly in their throats as did he. Then they were on the move; sixteen riders galloping from the castle over the draw bridge to meet the enemy. With their pages and squires riding behind, the men-at-arms moved, riding eight abreast, as they narrowed the distance between themselves and their adversaries. Eight visors clanked down into position and eight lances were lowered into the 'kill' position.

Sir Darwin's lance crashed heavily against the chest of the unfortunate who had drawn him to pair off with. Phillip and Maurice followed suit with their foes. Sir Jon's lance broke on the shield of his foe even as the other man's lance broke against Jon's own shield in the pass.

Pierre was amazed at the ease with which the young Duke unseated his rival. After the first pass, six of the eight challengers lay sprawled on the ground or in the pond itself. Realizing that they were completely outclassed, the challengers quickly regrouped and then galloped off in complete retreat. The victors filled the air with hearty laughter at the sight.

Pierre felt almost sorry to see the affair end so quickly. He found that he was so excited that he had not even found time to think of being afraid. Yes, he liked this kind of action.

"Well done" called out Duke Charles as he raised his visor. "To breakfast. This exercise has made me hungry."

CHAPTER THREE

Had Pierre been expecting a break in his training, he was sadly disappointed as there were months more of grueling training. It was more polishing of fiber, as it were. There was more tempering of the steel of his very being and more molding of boy into young man, who would one day become a man and one of the world's most successful soldiers.

Early the next year, war again broke out with the stubborn and pesky Marquis of Saluzzo. Once more, there was an abundance of violence, pillage and bloodshed. It began to get so out of hand that the Duke's juvenile but powerful cousin, Charles VIII, the King of France, had to step in and put a halt to the war. To let it continue would be to shed a bad light on the very crown of France and he could not let that happen. Too, the good Duke's image was beginning to dim within the Royal Court as some people thought the Duke to be too warlike and a bit too ambitious in some of his undertakings. There was even subtle talk of the young Duke, perhaps, coveting a higher crown. Duke Charles steadfastly maintained that he was merely trying to protect his Duchy. There were, however, hard feelings beginning to develop in Duke Charles toward his cousin, the King.

With war again put aside, the life of the court of Savoy settled back down to its usual festivity laden days. Pierre was privileged to attend a grand banquet in honor of a great German statesman, the Bishop of Sickau, who was personal ambassador of Emperor Maximillion.

From his temporary headquarters in Turin, Duke Charles took his troops on the road again. This time it was back to Moncrivello and then on to Vercelli. It was during this visit to Vercelli that Pierre first got to meet one of the most prominent of all statesmen of his time. This great personage was none other than the Regent of Milan, himself, the Italian Prince named Lodovico Sforza.

When Pierre had received word that he had been chosen as one of the four Pages of Honor at the great banquet given for the Prince, he was delighted. He was given a new suit of black velvet for the occasion. Pierre spent the entire evening watching the visiting Regent and taking note of all the things that went on during the course of the long and festive night.

Sforza was not a handsome man. His face was much too round and his Italian nose much too large and noticeably curved. His chin was far too ample and his lips remained too firmly closed. Yet, for all this, the great Prince demanded an air of respect. His quiet features held a special strength. It was reputed that he was a sensitive and intelligent man and was spoken of as being the craftiest diplomat of his time. He was said to be imaginative and emotional.

Still, he was a man who rarely lost his perspective or his temper. Lodovico Sforza was said to be a skeptical and superstitious man. Although a master of millions, the Prince was said to be a slave of his astrologer, not making a move without his consent.

During the evening's conversation, Pierre learned that the Prince had developed a vast experimental farm and cattle breeding station. Prince Sforza conducted experiments in cultivating rice, the vines and Mulberry trees. His dairies made butter and cheeses so excellently that Italy had never known the like before. Sforza's stables sheltered some of the finest stallions and mares in all of Europe. The great Prince had presented the duke with samples of his finest cheeses and made him a present, also, of two outstanding stallions.

More than once during the evening Prince Sforza noticed the hustling young Page from Bayard. Duke Charles was proud of Pierre and his rapid progress in all phases of his training and told the Prince so.

"Yon Page seems very devoted to his duties." remarked the Prince.

"Of a truth" replied the young Duke. "He is the best this court has ever had. His skill with weapons and horse are almost unbelievable. He is truly ahead of his years in both."

"Horse, you say?" inquired the Prince as his interest peeked, for his own love of horses was well known.

"Yes, he is an excellent rider. Would you like to meet him, Sire?"

The Duke motioned for Pierre to come to the table and the boy did so, without hesitation.

"Young Bayard," nodded the Duke, "our esteemed guest wishes a word."

Lodovico Sforza studied the tall, strapping youngster for a long moment then nodded toward Pierre.

"I am given to understand that you are a fair hand with a mount" said Prince Sforza with a broad smile. "I would very much like to see you ride for I am one who truly loves a good horse show. Are you really as good as they tell me, young Bayard?"

"The lad will be glad to demonstrate his skills for you, my Prince" said the Duke with a knowing smile.

Pierre blushed slightly and looked at the Duke,

who nodded and spoke quietly. "Ready your mount and we shall meet you in the courtyard presently."

Pierre lived up to his reputation by giving a fine performance for Prince Sforza, who was utterly and thoroughly pleased. The Italian was so impressed that he gave Pierre five crowns of gold for his purse. He even made a subtle attempt to steal the young Page away from Duke Charles by offering to let him travel with him and ride for some of the crown heads of Italy.

Of course, the Duke of Savoy flatly refused to part with the young man from Bayard. It was an experience that would stick with the young Page for a long time, though.

The sun beamed down on her silken brown hair, lending a halo appearance to her soft, refined face. She walked with a bounce of happiness in her steps and her face beamed as she spoke with two other maidens of the court who walked with her.

Pierre found himself fascinated by her simple beauty which seemed to reach out and grab his entire being. Where had she come from? Who was she? Why had he not seen her before? The questions buzzed through his head in a never ceasing stream. The young Page felt a bit odd as he had never before taken such interest in any female. Yet, here he was, almost mesmerized by this young beauty.

As the three women passed him, Pierre found his gaze meeting the young girl's own gaze for the briefest of moments. Yet, it seemed like a lifetime as he drank in every aspect of those soft brown eyes that danced and twinkled in the afternoon light.

He smiled and she slowed her steps just a bit before returning his gesture with a quick little smile of her own. She bowed her head, ever so slightly, in recognition of him, and then she continued on her way. Just before she and the other ladies rounded the corner of the building, however, she turned to steal another look at him and offered another small smile.

Pierre felt the warmth which consumed him. He quickly glanced around him to make sure that none had seen how he was behaving at seeing the girl. Nobody was paying him any attention and, for that, he was grateful. However, he was so impressed by her that he, immediately, sat out to discover just who this beautiful young girl of about his own age was. He had not had time or the inclination to pursue a female before as his life was much too full to let anything or anyone turn him from his demanding training. Yet, something about this young woman caught his attention as much as if someone had hit him in the head with a board.

He soon discovered that the girl had recently joined the court of Savoy as one of the Duchess' ladies in waiting. She was from Grenoble, only thirty miles from his own home. She was the daughter of a high ranking diplomat of Grenoble.

Her name was Anne de la Chavet and she was only six months older than he was.

During the next month, Pierre saw Anne de la Chavet on several occasions. He was flattered a bit and somewhat amused to have discovered that she had been making certain inquiries about him, as well. He was pleased at this. She actually seemed to be looking for him in the courtyard whenever she was out and about on the Duchess' business.

Each time they met or saw each other from afar, they exchanged courteous greetings or small smiles. Their respective schedules, however, kept them from having much time to get really acquainted.

Then it happened. He finally got the opportunity to meet Lady Anne and his young heart almost burst with joy when it happened.

He had just given a riding exhibition in honoring a French Cardinal named George D'Ambois, who was visiting Savoy. Pierre was getting a drink of water from the big well when, out of the corner of his eye, he saw her coming toward him.

As she approached, he wiped his mouth and quickly dipped the ladle back into the water bucket. He politely offered her a drink. He got the feeling that Anne really did not want a drink but, rather, took the offered opportunity as a perfect chance to finally get to meet this tall young man of the court.

Pierre watched her delicately sip from the ladle. Their gazes locked into one another as she handed the wooden ladle back to him. "Thank you for your kindness" she said quietly.

As he took the ladle from her, Pierre's fingers lightly rubbed against hers. He instinctively started to jerk his hand away as he did not want to appear, in any way, ungentlemanly. However, her fingers stopped, for the briefest of seconds, on his as she smiled demurely.

"You truly are a most magnificent horseman" she said softly.

Pierre felt his cheeks flush a bit and purposely avoided her intent gaze.

"Would it please you to learn that I have found out a few things about you, Pierre Bayard?"

This, he had not expected and he was taken completely off guard. He hoped that she had not notice his astonished look and think of him as a complete dunce. He recovered quickly by pretending not to be surprised.

"I trust that it was not too boring of a research." he heard himself say and could have kicked himself the moment the words left his mouth. A gentleman did not speak to a lady, especially a lady he did not even know, in that fashion. If she was offended she did not let on.

"No, not in the least" she answered matter of factly. "In fact, I find you to be quite interesting. For a boy, that is."

"You are much too kind to say that, my lady" he heard himself mumble softly. Pierre was pretty sure that she caught his 'double take' on the words "for a boy".

"Forgive me, but my lady scarcely is of any great age, herself."

Anne smiled slightly "I guess I deserved that one".

"Did you really like my riding?" asked Pierre, thinking it might be the more prudent thing to do to change the subject altogether.

"Of a truth, yes" she was quick to answer. "And I do hope that you don't think me too bold for talking to you this way."

"To be sure, I am pleased" he said. "I must confess that I have wanted to meet you for some time and was hoping you did not take me to be a lout."

"Well, since this seems to be a time of confession" she said with a pleasant smile. "I must confess that I, too, have been trying to meet you. However, your training schedule seems to consume all your time."

"I fear that is true" he said with a small sigh. "But it is unavoidable as to become a knight is the one thing I wish of my life more than any other."

"I know you already know, but I am Anne de la Chavet" she said with a small giggle and a slight curtsy.

"Pierre Bayard." he countered as he could already feel himself becoming much more at ease with her. She was easy to talk to. Not, at all, as he would have suspected a beautiful woman to be.

"I must go now" she said as she looked around. "I mean before some wagging tongues make me out to be a brazen woman or something. I just wanted to say something to you while I had the chance." Anne looked into his grey eyes and added with a small tease in her voice. "If a girl waited on you, Pierre Bayard, she could grow to be an old lady before it happened."

Before Pierre could even respond, Anne turned and hurried away.

"Oh! I promise you will see much more of me, my lady" he said quietly in a voice barely audible.

Soon after Pierre and Anne had officially met, the House of Savoy packed up and was on the move again. This time it was a trip to Nice where there were great tournaments and feasts. Pierre received several chances to display his riding skills as Duke Charles seemed always to delight in showing him off.

During this time, too, Pierre received a special treat. There was a Squire's tournament held for the fun and pleasure of all. Pierre was allowed to enter, as he had proven, time and again, to be one of the best Savoy had to offer. Pierre outdid himself as he unseated ten straight challengers, including Henri, and was crowned the winner.

By the time his fourteenth birthday came, Pierre was a tall, strapping young man of extremely fine proportions. Standing five feet and ten inches tall and weighing one hundred and seventy pounds he could hold his own quite well against most of the other Pages and Squires. His whimsical smile seemed to captivate the hearts of all who knew him.

All the while, Pierre and Anne seemed to be growing closer and closer. Even though he was the topic of talk among all the ladies of the court, none but Anne seemed to mean a thing to him.

All was not festivities, however, as the Duke's popularity had seemed to grow less than grand in the higher courts of France. His continuous open warfare with the Marquis of Saluzzo, who was much too much of a Courtier to be banished by the king, seemed to be turning the tide against Duke Charles. There even was heard bickering and restlessness within the Duke's own household.

    Pierre, himself, never understood nor cared much for politics and, thus, never delved very deeply into what was actually taking place within the Savoy household. The young man from Bayard, by choice, became oblivious to the bitter rivalry and undermining of positions that was going on in the inner structure of the House of Savoy. Thus, Pierre was much surprised when it all blew up one day as Duke Charles discovered and put an immediate and permanent end to some particularly vicious dealings of some traitorous acts within his domain.

    Young Pierre found that the results of the Duke's investigation involved the banishment of his own cousins, Antoine and Hughes de la Forest and the dismissal of Guillaume from the ranks of the Pages. Pierre, himself, though remained completely unscathed by whatever the problems that his kinfolk had caused and suffered no ill feelings from the Duke or from any of his household. Pierre was a well-liked young man by all and remained the favorite of Duke Charles. The boy from Bayard was considered the finest Page in the history of the Duke's service. Duke Charles favored Pierre with almost the same affection that he would have held for a younger brother, had he had one.

In early March of 1489, Pierre was summoned to an audience with Duke Charles and Duchess Blanche. He was puzzled when he received word that the Duke had sent for him. Pierre checked his appearance carefully before entering the great hall, where his hosts were waiting.

"Your Graces?" he said, as he fell to one knee and bowed.

"Please, arise young Bayard." bid Duke Charles.

Pierre did as he was bid, slightly perplexed at the entire setting.

"It is with great pleasure that I tell you of some good news in your person" said the young Duke as he seemed to almost burst with pride of what he was trying to relate to Pierre. Duke Charles cleared his throat and glanced over at his lovely young wife then looked straight into Pierre's steely grey eyes.

"I have received, this day, from my cousin Charles, King of France, a personal invitation to go to His Majesties Court." announced the youthful monarch proudly.

"We will be leaving within the week. There are many matters to be settled between His Majesty and myself. Ah, but I did not summon you before me to discuss politics."

Duke Charles exchanged knowingly excited glances with his beautiful wife and then continued. "I will have you accompany me wherein you will find yourself heaped with no small honor in your own right. You will ride in a Command Performance before His Majesty at the Grand Tournament, which is the largest and most prestigious of the year."

The Duke waved a letter in the air. "This is your passport, young Bayard" continued the enthusiastic duke. "It is from His Majesty's Regent, his sister Anne De Beaujeu, who bids you welcome and says that King Charles would be honored to see you ride. Young Bayard, this is the first time in the history of any country that a page has received a special invitation from a king. We are most proud of you. Most proud, indeed!"

There was a moment of dead silence as Pierre tried to fathom all that the young Duke had just told him. Pierre felt himself fight back the great tears of joy. The King of France had requested that he ride for him. It was an honor befitting even the most Royal Princes of the Blood. Surely the news would reach back to Castle Bayard where his parents would hear of the event.

"I will be honored to ride for His Majesty." Pierre said in a barely audible voice as he was still grappling with the entire conversation.

Within the hour the grand news had spread throughout Savoy like an unchecked blaze in the forest. Pierre was the hero of the hour. The boy from Bayard felt, at the same time, deeply gratified and humbly frightened at the large demand put upon him. Pierre knew that the Duke and his cousin, the king, were experiencing conflicting views concerning many matters and their feelings had been strained. His fond hope was that he would not be hurt in the fall out.

The subject of Saluzzo was, indeed, a kindling point. Saluzzo was under the protection of the French Crown and Duke Charles had openly shown great disdain toward the Italian, which was not pleasing to the French court. Yet, King Charles had sent for the Duke. So, reasoned Pierre, the rift between them surely was healing. At any rate, no trouble was expected as the Duke would want to put on his best face for his cousin. Duke Charles stood on precarious ground to be sure but could still gain favor from the French Court if things went well. Pierre realized that his small riding contribution could fit into the Duke's plans of doing just that. If what he did could help the Duke then Pierre was more than happy to do it.

It was with all this on his mind that Pierre went to the corral and called Wingfoot to him. As he stood rubbing the great horse's mane, Pierre talked softly to the animal. "Well, my esteemed mount, I guess you have heard the news by now." he said aloud.

"We are going to the Royal Court and ride for the King, himself. I will admit that I am a little frightened, good horse."

"No need of fear, mon ami."

The friendly voice of Jacque spoke from behind him. "The Duke is right, you know?"

Pierre turned to face the smith, who was approaching with a huge smile on his face. "I doubt that there is a finer pure horseman in all of France. We will be counting on you to bring fame to your name as well as to the House of Savoy. I know you can do it, lad. You have all the markings of a fine knight. Someday, you will be a glorious man-at-arms. People will remember your name and tell stories of your deeds for hundreds of years after you leave this earth."

"You are a good friend, Jacque" said Pierre, who now stood nearly as tall as the smith. "I know you mean to cheer me, but I am still nervous."

"Nervousness is a good thing in some cases." pointed out Jacque. "You will do fine. But, even if you do not, remember that I, Jacque, will always count it an honor to be the friend of Pierre Bayard."

It was two full days later, near dusk, when Pierre received word that he was to join the huge party of over 1,400 travelers, which included almost everyone in Duke Charles's service, in the beginning of the trip from Chambrey to the Royal Court at Amboise.

With a lot to think about, and feeling a trifle nervous, Pierre decided to take a quiet walk in the Duke's flower garden. As he strolled slowly among the flowers, Pierre became aware of being watched.

Immediately, he spotted the form of the woman walking toward him. She was half hidden in the shadows but Pierre knew it was Anne simply by her step.

"I heard that you were walking here." she said quietly as she approached.

"I wanted to see you."

"See me, my lady?" he playfully replied as she stopped before him.

Pierre had longed to see her as much as she wanted to share his company. With such rigorous training, he had not been given much time for socializing.

"I... well, I wanted to tell you how very proud I am of you. How proud we all are of you."

"That does please me" he answered softly. "I only hope that I can do all that is expected of me."

"Oh, you will, Pierre" cooed Anne as she reached out and touched his hand. "And much more. I am sure of it."

No words came from him, even though he wanted to say something to her. She seemed to have that kind of an effect on him. He wasn't sure why, but he always felt a little nervous around her.

Detecting his loss for words, Anne decided to ease the moment. "I do hope that you do not think me too bold for what I wish to ask of you" she said softly.

"Ask anything" he quickly said. "If it is within my power to give, it is yours."

"I beg you to accept a token of great feeling from me" she said as her hand dipped in the pocket of her dress and pulled out a beautiful, folded green and white silken scarf. She gently laid it in his hand.

"I truly could think of no greater honor, my lady." he said.

They stood gazing into each other's eyes in the electrifying moment as she gently reached and took his strong suntanned hand in her own. Sensing that she wanted the moment to last as much as he himself did, Pierre cupped his other hand over hers. "It is a wonderful gift" he said.

"Would you wear it as my favor upon your sleeve when you ride for His Majesty?"

"Bon Dieu!" he almost shouted. "I will be proud to wear it. I will keep it always."

Anne stood looking up at him in silence for a long moment. Her expression was one of almost sorrow. When Pierre detected this he put a finger under her chin and lifted it slightly until they were gazing fully into each other's eyes.

"What is it? You seem so... sad" he asked.

"If... If His Majesty really likes you... and wants you to remain with him...?" she began, then let her voice trail off in the unfinished sentence. A single tear escaped down her lovely face as she quickly turned her head away in hopes that he had not noticed.

But Pierre did notice and as he started to say something to comfort her, she was already speaking again. "If this thing would happen. Would you... we ever see each other again?"

Pierre reached out and took her by the shoulders and turned her to fully face him as he searched her eyes with his penetrating grey eyed gaze.

"Would it matter if we did not?"

As soon as the words were out, Pierre regretted saying them. It was a foolish thing to say. Of course, it would matter. They were in love and each of them knew it.

Anne stretched to kiss him on the cheek. "Of course, it would matter. Don't you know that?"

"I know that" he said softly with a nod.

Then, following a pure impulse while gazing at her soft red lips so near to his, he acted as any young man would act toward the woman he really cared for. He bent down, slightly, and placed a soft kiss on her lips.

For an instant, she responded by pressing her mouth hard against his as she clung to him for a desperate special moment.

Then Anne was gone, hurrying away so that he could not see the tears of joy.

Pierre stood silently gazing at the spot where she had last been seen. He looked down at the neatly folded green scarf in his hand. Pierre felt really good inside. Anne was his first love.

Sometime later, in his quarters, as he wrote his father of the upcoming journey, Pierre still felt the warmth of her softness on his body.

As the journey began, everyone knew it would be a long trip. Pierre began to understand just how long as the duke's party made its way up through Belley and on to the great city of Lyons. Then it was on to Bourges and a big swing over to Tours, where they spent nearly an entire month in continual games, banquets and other festivities. Pierre's birthday was in April and he turned fifteen there in Tours.

During these times, it became more and more evident that Lady Anne was the lady love of young Pierre Bayard's life. Their eyes spoke volumes with each look. Even the knights would smile with pride as they remembered, no doubt, their own first experiences with loving a girl. They seemed especially pleased over the fact that this encounter involved the Court of Savoy's very favorite Page. As such, Pierre soon discovered that the people seemed to take an interest in just about every aspect of his life. That alone was a bit mind boggling to the young Page.

However, the daily talk was not all centered around Savoy's favorite Page, It turned, too, to the great tournament ahead at Amboise. Sir Darwin and Sir Phillip would be representing the House of Savoy against some of the finest knights in all of Europe in the greatest tournament in the land. It promised to be a very exciting time and, whether any wished to admit it or not, a shrewd political game of one-upsmanship between many of the parties.

From Tours, the party made its way on through Plessisles-Tours and then found themselves on the last leg of their journey. After a short stay at the Royal Court, Duke Charles was planning to continue on to Chateau Renault and Vendome, where he had more relatives and more business to attend too.

Pierre was riding, as he had been throughout the journey, directly behind Anne's party when he first caught sight of them. They were most impressive.

Those two dozen brilliantly garbed riders, who crested the hill and guided their mounts straight toward the Duke's party.

Each rider was a battle hardened knight in the King's own service. They made quite a sight in their various types of helmets, their glistening chain mail and armor and their cleverly decorated shields as they carried their banner adorned lances high. Each knight, too, proudly wore his gleaming golden spurs, which had been given as a signification of him obtaining knighthood.

Duke Charles and Sir Darwin greeted the large escort and thanked them for meeting their party. In a generous display of hospitality, the knights bid the new comers welcome in the king's name. Then the knights wheeled their great mounts and fell in line, half on one side and half on the other, with the travelers to accompany them the remaining distance to Amboise and the Royal Court of France.

Swinging his horse in to ride beside Anne was a bold young knight carrying a Crimson Shield with a blue dagger on it. Pierre, quite uncharacteristically, felt himself grow more than a little annoyed at the man as he watched him cast a bold gaze upon the beauty from Savoy. The knight removed his helmet to reveal a bearded face and a crop of long black hair. His smile told Pierre that the man considered himself quite the ladies' man. He was acting a bit too cock sure of himself for his own good.

"My lady." said the knight as he bowed slightly in his saddle to her. "This gentleman would be most greatly honored to wear the colors of such a beauty as yourself in the great tournament to be held four days hence."

"Sir Knight" answered Anne with more boldness than Pierre had expected of her. "I am sure your offer is well meant and would truly compliment many a fine lady. However, I have already favored a champion of France. Behind me sits he who is the greatest horseman of the land. He will be giving a command performance for His Majesty, himself, at the great tournament. I thank you for the generosity of your offer."

At Anne's words, the knight with the Crimson Shield turned to look straight at Pierre. His gaze was one of confusion as he said nothing to the young man from Bayard, but turned back to face Anne.

"But, my lady." he said in obvious disbelief. "He is but a lad and a mere Page at that."

Pierre felt himself bristle at the words of this man who was barely out of his teens himself. The Page from Bayard battled within himself that he might not do anything foolish, which might disgrace himself or Duke Charles or Anne. This man might act rude and brash but he was still a knight in the

King's service and Pierre was only a lowly Page. Still the words stung and were unsettling to the boy.

Lady Anne, however, was bound by no such rules. She recognized that Pierre could not respond in his own defense without getting into trouble. She answered, quite nicely, on Pierre's behalf.
"A page, yes, Sir Knight." she spoke up proudly. "But he is my champion above all others. He will be the one wearing my colors in the tournament."
Anne's words seemed to put the young knight in his place for the moment. He bowed slightly, looked back at Pierre, once more, then replaced his helmet and rode the rest of the way to Ambroise in complete silence.

The tournament site was a beehive of activity with hundreds of people busying themselves with preparations. There were old friends greeting each other and new friends making acquaintances. There were tents and pavilions and awnings erected everywhere. The dozens upon dozens of tents and pavilions spread over the giant field resembled lines of carefully cultivated crops. Over each entrance was painted a coat of arms, signifying the identity of the knight in residence. In front of each tent or pavilion stood the particular knight's banner and lance and, in some cases, a mounted shield standing proudly at the guard.

Everywhere, groups of people gathered and were talking and joking and laughing and singing. Some even danced. They spoke of other tournaments past and of old castles and new ones being built and of battles survived and anticipation of battles to come. Many chose to lay wagers on who they considered the favorite to win this magnificent event. Each time word of some new gallant's arrival passed on to the group of wage makers, they would reshuffle their thinking and betting.

Pierre, like everyone else in the Duke's party, took in all aspects of what was happening with excitement and much anticipation.

The escort led the Duke's party to their designated area and then took Duke Charles and some of his special entourage on to a great wing of the Royal Palace.

Pierre had never been to a court of a king before. For that matter, he had never even seen a king before. So, he tried not to display the obvious awe that he found himself experiencing.

After tending to the chores given him, Pierre went to the stables to tend to Wingfoot. He had just finished giving the animal a well deserved grooming when he spotted the lovely pair at the stable entrance.

"Your animal is truly a beauty" said Anne as she coaxed her friend to come closer to Pierre.

"Thank you." said Pierre as he looked around nervously. "My lady, you two should not be here."

"Then take us away" retorted Anne playfully. "Would you be so kind as to escort us to watch some of the trial jousts?"

Obviously uncomfortable at having the women in the stables, Pierre nodded quickly and led the way out.

"This is my friend, Claudia d'Penzant." introduced Anne.

Pierre bowed slightly toward the other woman and then took Anne's arm as they walked.

Some of the knights were already holding trial jousts, warming up and making final adjustments for the tournament. Their games mostly were of light sword play and archery contests.

There were a dozen or so men in the largest of these archery groups when Pierre and the ladies arrived. Pierre spotted Sir Darwin who was one of the bowmen. He and one of the king's archers were locked in a particularly close contest. A tall man, clad in black with a thin black veil across his face, stood a close third.

"He carries himself particularly well" observed Pierre to no one in particular. "I wonder where he is from."

A quiet voice just behind him answered the question.

"Indeed" spoke up the tall gentleman standing just behind the trio of newcomers. "He is an Italian Count, or so I have heard. I know not his name as it is rumored that he prefers to be nameless for some political reason or another."

Pierre and the women nodded slightly in appreciation of the insight.

Pierre felt that this would be a man to watch closely in the tournament when things counted for real. There was just something 'special' looking about him.

From the archery field the three wanderers moved to another venue where more men were going gamely at each other with wooden sword play. Sir Phillip and a large man were hooked up in a spirited contest. The knight from Savoy left himself open for only an instant and paid the price. The big man won a good point with a healthy swat across Sir Phillip's unguarded belly. Pierre hated to think of the death of Sir Phillip had the battle been for real with real swords.

Suddenly, Pierre became aware of the crowd, which had grown strangely quiet as everyone seemed to be straining to get a look at someone standing at the end of the lists. Someone near Pierre whispered a name. "Sir Claude, himself."

"Who is it, Pierre?" asked Anne as she tried to see over the shoulders of some of the onlookers just as the man turned away and disappeared behind one of the tents.

"Who is it, my lady?" asked a rotund man dressed in expensive robes who had overheard her question. "That is none other than Claude de Vaudrey. He is the finest swordsman in all of France. He has not lost a sword match in over three years. He is somewhat of a hero of France me thinks."

"You mean another hero of France" corrected Anne playfully as she beamed proudly in Pierre's direction.

"My lady?" said the man with a puzzled look.

"Nothing, sir." replied Anne. "Thank you for your information."

Pierre thought that Anne did not understand the significance of Claude de Vaudrey's position in life. But, then women, with the exception of his personal heroine, the warrior maid, Joan, rarely understood men and their world.

Continuing on, Pierre and the two ladies came across a group of people watching more jousting. This time, however, it was a group of squires who were engaged in a spirited wrestling contest. Henri de la Palu de Varembon was rejoicing in being declared the victor.

"Seven straight foes. What a nice feat" remarked someone nearby.

"That means you could have been the champion had you entered" remarked Anne as she looked at Pierre with no small pride in her voice.

"You are much too kind." responded Pierre. "But they are squires."

"So they are. But you have defeated Henri on several prior occasions." countered the beauty "You, sir, are much too modest."

Deep down, Pierre did feel a certain amount of pride as he knew that what she said was true. He had to be careful not to let himself get too swelled with pride, though. Feeling a bit uncomfortable at the praise being ushered toward him, Pierre quickly looked for relief and found it in his next sentence.

"Come, I will take you back" he said. "I have more chores to attend too."

CHAPTER FOUR

The summer sun had already begun its rapid descent when Pierre, wearing one of his finest grey and red tunics, walked anxiously toward the big blue tent. He could feel the excitement build with each step that he took. Outside the door way of the tent stood a pole that supported the very familiar Crest of Bayard.

Pierre did not know exactly why he hesitated before the closed flap, even after he had gained permission from the guard to enter when he convinced the man who he was. It seemed ages since he set eyes on any of his family. Pierre guessed that it was just nerves that slowed his steps.

Inside, he could hear the voice of his father as he called out to someone. Then he heard his mother's voice answer and a warm feeling flushed over him. Pierre could also hear the voices of his sister, Jeanne and his brothers' George and Jacques. After making a final check of his appearance, Pierre took a deep breath and pushed back the hanging flap and stepped inside.

Suddenly, all the movement inside the tent ceased as every gaze fell upon the tall, strapping young Page from Savoy. It seemed that time had actually stopped as he waited for recognition to set into their minds.

Old Lord Bayard stared then whispered "How much like his grandfather he looks."

Lady Bayard and Jeanne rushed to shower Pierre with hugs and kisses and his brothers gathered around, touching his arm and patting him on the back. Lady Helen called out his name again as she wiped away tears of happiness.

Pierre could feel a couple of his own tears roll down his suntanned cheeks as he hugged his mother and sister then stepped away from them.

"My honored father" he said quietly as he watched the old man rise slowly from his chair.

Then father and son hugged. It was a special moment, one that Pierre had wondered if it might ever happen, after he left his home years ago.

"You look well, my son" said Lord Bayard, finally.

"Thank you, father" said Pierre. "They have been looking after me very closely."

Pierre looked around and then asked. "Where are the others? Where are Phillip and the girls?"

"Your little brother fell from his horse and broke a leg" answered Lady Helen. "The girls stayed home as the travel might have been too much for them."

"Phillip never could set a horse" said Pierre with

a smile.

"Certainly not like you" answered Jacques.

"Nobody could set a horse like him" laughed George.

Lady Helen then took her son by the hand and looked deeply into his eyes.

"We are all so very proud of you, my son. You will gain great favor with His Majesty when he sees you ride."

"Perhaps, my mother" agreed Pierre. "However, I must confess of my nervousness over all the excitement of it. What if I do badly?"

"Do not dwell on it, boy" said Lord Bayard. "You are one of the finest horsemen alive. You will do the houses of Bayard and Savoy proud." The old man nodded a reassuring nod then changed the subject. "I hear that Duke Charles has a very fine champion within his walls."

"Yes, sir" answered Pierre. "He is called Darwin and he is as good a swordsman as I have seen. Some even wager that he will be the one to take Sir Claude."

"That would take some doing for any man" admitted Pierre's father. "I will look forward to seeing the outcome. As for us, we brought Merrick. He is a sturdy young man who came to us from your uncle, who is now the new Abbe of Ainey. He sends you good wishes, by the way."

"Enough man talk" ordered Lady Helen. "Tell us of yourself, my son."

Pierre spent some time trying to relate all the happenings he had experienced since coming to Savoy. His audience listened with eager captivity until the hour grew very late. Finally, Lord Bayard called a halt to the visit and suggested they all retire as the morrow would bring another long day.

Pierre was very contented as he left his father's tent and made his way back to his own quarters. It might be a long while before he could see his family again. He had no way of knowing. This, though, was a fruitful and pleasant visit and nothing nor anyone could take that away from him.

On the way to his quarters, Pierre passed one huge tent which had gay music and hearty laughter coming from it. Pierre saw many banners hanging from it and knew this was where the King's own knights dwelled. With the aid of the flickering light from the camp fire, Pierre was able to spot the Crimson Shield with the blue dagger.

At that exact moment, the tent flap opened and four women came scurrying out. They were giggling and adjusting their clothing while holding money in their hands. Pierre watched them disappear into the darkness. He gave the tent a final look and went to find his own bed. The young Page from Bayard had an uneasy feeling that, somehow, the knight with the Crimson Shield and blue dagger would be no small part of his upcoming life. He tried to dismiss it from his tired mind but that thoughts stayed, annoyingly, in the background as he continued to make his way toward his own quarters.

Many thoughts filtered through Pierre's mind as he lay on his bed for a long time, too excited to drop off to sleep immediately. The last thing that the boy from Bayard did remember before sleep overtook him was the green and white silk scarf that he clutched in the fingers of his left hand.

As the first splintering fingers of dawn eagerly scratched at the morning sky, the vast area of tenting was no longer a village of people on a merry holiday. Instead, it had transformed itself into a village of dedicated people who were extremely hard at work. Not one single person was idle this day as each had an important task to be completed.
Some tested the handles of their particular champion's shields, making absolutely sure that it would withstand the rigors of a sword or lance attack. Others made certain that the shining armor had its final coat of paint. The rivets, buckles and straps had to all pass final examination, as well. The horses, too, were given extraordinary care this day.

They were fed and rubbed down after their morning exercising. Every aspect of their rigging was carefully inspected for the slightest flaw.

Every tiny portion of the great household was just as meticulously scrutinized from the kitchen to the sewing room and from the stables to the armory. Even the musicians gave more than their usual special attention to checking their clarions, trumpets, kettle drums and pipes.

To all, the success of the entire day would depend on how well each performed his or her given task.

The topic of conversation was also altered and confined itself strictly to the activities of the day. Each champion had his own special merits and skills. Of course, too, every lady had her own favorite knight carrying her particular colors into the games.

With all the great knights present, the talk of the games seemed to somehow come around to the beautiful young lady from Savoy who had given her favors to a mere Page when she could have had her pick of any gallant in the field.

Just as importantly, they all seemed to be aware of the fact that this Page from Savoy by way of Castle Bayard, was no ordinary young man and all awaited to see just what was so extraordinary about him. The interest was even keener when they learned that Pierre was, himself, the son of one of the countries most respected and honored men-at-arms.

Being the official umpires of the games, the ladies were honor bound to be fair. They had the final say in any dispute of the proper awarding of the prizes.

While their attendants were busying themselves with the outer preparations, the contestants staunchly spent their time preparing their inner selves by attending mass for more than an hour during the early morning hours.

With the contestants emerging from their various places of worship, everyone turned their attentions toward the lists with eager anticipation of great contests to be held.

The lists, themselves, had been prepared hours in advance. They consisted of a vast level area which had been fenced off with a double row of wooden railings. Between the railings was saved a space for those who were to assist the injured knights during the tournament. On either side of the rows, erected at a safe distance, were huge galleries, handsomely decorated with banners and tapestries and enhanced by the brightly colored dresses and gowns of the lovely ladies.

King Charles had already announced that lances and broadswords were the only weapons to be used in the great tournament. As suspected, all weapons with sharp points were banned. The points of the lances were removed or otherwise protected by rockets or by coronels.

Pierre had to admit that he was more than a little awed when he witnessed King Charles and his huge entourage enter the grounds and take their places. He had never seen a king up close before and it was quite exciting.

The King's heralds proclaimed the rules of the day. Of course, every contestant already knew them by heart. They were well aware that any contestant who broke the most lances was to have first prize. Naturally, the lances had to be broken according to the rules. If a knight broke a lance while striking an opponent from the saddle, he would receive three points for his feat. A loss of a point would be assessed against the man who broke a lance on an opponent's saddle. To meet an opponent's coronel twice with your own would gain a point.

However, this feat was not held above the unseating of an opponent. Any contestant who struck a foe's mount would automatically be out of the contest. The same would be true of any knight who would strike an opponent while his back was turned or while he was unarmed. To break a lance across an opponent's chest was deemed a most shameful act as it showed poor riding and poor aim and would result in a loss of two points. If any opponent were to remove his helmet for a rest or a breath of fresh air, he was not to be touched until the helmet was back in place.

With the rules given, the chief constable then examined all the weapons, the saddles and their fastenings and ornamentation to insure that no rider could stick to his horse in any manner other than with good riding skills. When the constable was satisfied that all was in good and proper order he nodded to the heralds.

The exuberant heralds turned to the waiting contestants and proudly cried out in loud voices. "Come forth, knights. Come ye forward."

A nervous Pierre Bayard looked over at King Charles' box where the monarch sat with his sister, Anne De Beaujeu and twenty beautiful ladies of the court.

King Charles, himself, was but nineteen or twenty. Charles was not a handsome man. He was sickly looking and a small man who possessed an almost grotesque parrot nose. Yet, Pierre cared not about judging a person by their physical appearance. This man was the King of France and what a glorious and serious position he did hold. Even though his sister ran the country until Charles was deemed old enough, he was still the King and he deserved all the respect that went with the title.

The lengthy cavalcade of beautiful horses and handsomely clad knights then began its slow procession before the king's box. Each horse was led by a beautiful lady whose colors that particular rider was wearing. Each rider also passed by the Arch-Bishop Georges d'Amboise and the Cardinal Antoine Du Prat and then on before some of the mightier lords and soldiery of all France. Pierre was very proud of the fact that, among these, sat his own family.

Near King Charles sat old Marshal d'Esquerdes, Phillip de Cre'vecoeur, the chiefest warrior of France. There, too, was Marshal de Gie and Marshal de la Tremouille. Count d' Ligny, Matthew of Bourbon and Berault d'Aubigny were there in all of their royal splendor as well as Genouillac and countless others who were the very cream of the French Cavalry.

Pierre was particularly pleased to see a young captain that he had heard many things about and a man whom would become his personal favorite in the future. Young Captain Louis d`Ars, a very accomplished warrior beyond his years was astride his big chestnut colored battle charger talking to his squire when Pierre fist saw him.

For the briefest of moments their eyes met and Pierre quickly looked away as he didn't want to have the man think he was staring.

All the knights were presented in pairs with each pair being separated by a robe. The beautiful ladies returned to their places in the galleries after their champion had been presented. Each knight made his salute to the king and then rode to their pre-destined groups. Behind them rode their proud squires, sometimes as many as three to a knight.

When each was in place, King Charles nodded the signal to the ladies, who, in unison, cried out "Let the games begin."

With trumpets screaming their shrill songs into the morning sky, the heralds bid the play begin with a shout: "Do your duty, valiant knights!"

The ropes at each end of the lists were snatched away to a deafening roar from the spectators.

Riding three abreast, each knight called out the name of his lady love.

From the galleries came the happy cry of "Onward brave knights. The eyes of the beautiful behold you."

As they rode, too, each knight bent low in his saddle while holding his lance at the 'rest' position. The first six contestants raced forward, each in his respective lane, as the minstrels played and the trumpets kept up their ear splitting blasts. At the last possible second, each lance was lowered into the 'kill' position. The next moment was a thrilling mixture of sounds; the crackling of ashen lances, the trampling of horses hooves, the spontaneous cheers of the crowd and the agonizing moans and groans of the fallen and injured.

Only two of the first six riders emerged unscathed from the confrontation and still upright in their saddles. These two were successful in unseating their rivals while the other four all managed to unseat each other.

These two victors gave a triumphant wave to their respective ladies and, with the crowd cheering wildly, went to the space that the king designated as the winners' area. There, they would check their equipment and rest and make ready to meet whoever was their next opponent.

The lists were very long. Wave after wave of contestants rode at each other with the winners gaining the right to engage in another round on the following day. Hundreds upon hundreds rode with some falling and some being injured and some going to the winners' area to wait until the morrow.

Pierre Bayard watched with total infatuation. He followed the progress of some favorites but concentrated on two or three in particular. In the tenth wave, the Black Knight from Italy unseated his foe with one, easy and powerful stroke of his stout ashen lance. The appreciative crowd applauded vigorously with enthusiasm.

Captain d'Ars, one of the court favorites, easily unseated his foe to move on to the next round. Two more lucky victors of the day were Lord Bayard's champion, Merrick and also Phillip of Savoy.

It had been an exhausting day for all but, to be sure, also one filled with excitement and memorable events that they would talk of for months to come.

Dusk was falling when the final six contestants took their places in the lists. On one side were three fine challengers. On the opposite side was Sir Darwin of Savoy, who was flanked by two of King Charle's finest. It both irked and surprised Pierre to see the markings on one shield which was the Crimson Shield with the blue dagger. On Darwin's right rode a man whom evoked a thunderous applause from the weary gallery; Sir Claude de Vaudrey.

Sir Darwin greeted Sir Claude with a small nod and then let his visor fall into place.

In unison, with their magnificent horses almost matching strides, the three valiant men rode and their lances dropped into the 'kill' position. The air was permeated with the sounds of lances on shields and lances against lances. Three victims fell and three victors rode proudly to the winners' area.

Thus, Sir Darwin, Sir James and Sir Claude all joined the other victors of the day. The next day all the winners would pair off and continue until only two would be left to fight for the top prize. Tomorrow, also, the tournament would choose its 'Queen of Love and Beauty' from among the bevy of gorgeous ladies in the gallery. Last, but by no means least, Pierre Bayard would also have his day. He and Wingfoot would ride for the King and all of France.

Sir Phillip was jerked violently as the deadly accuracy of the Black Knight's lance slammed against his shield. For a brief moment it appeared that the Savoyan would recover to make another pass. But, he finally toppled from his grey battle charger and fell in a clanking heap upon the hard ground. Pierre was relieved to see Phillip sit up and remove his helmet. His ego was bruised but nothing more.

The Black Knight from Italy had vanquished his third foe of the day.

It was at this time that King Charles, following his whimsical ways, called for an interruption in the jousting and announced that it was time to select the 'Queen of Love and Beauty'.

The nominees were announced and paraded before the gallery and the ballots were cast. Pierre was stunned when he heard the heralds loudly announce that the tournament's 'Queen of Love and Beauty' was Anne de la Chavet of Savoy. Pierre was so very happy and proud. Anne was beside herself but took it all in stride.

The next announcement should not have come as such a surprise to the young man from Bayard. It was the biggest reason that he was here in the first place. Yet, when it was actually his time to bask in the limelight and be the focal point of everyone's attention, Pierre felt himself gulp loudly and tried to quell the burst of nervousness that beset him.

More excitement enthralled the galleries as the young Page rode boldly out and reined in Wingfoot before the Royal Box. He sat proudly on his horse with Anne's green and white scarf secured tightly around the sleeve of his bright red and grey tunic. Removing his cap, Pierre guided his beautiful horse into a bow, which had taken him weeks to teach the animal how to do. This seemed to greatly please all in the Royal Box.

The young monarch looked deeply into Pierre's eyes as he spoke.

"To have France represented by such glorious young men such as you, Pierre of Bayard, is a fine tribute to her future. Ride well young Page. All of

France is watching you this day."

"Your Majesty!" said Pierre with another bow. Then he returned his cap to his head, backed Wingfoot up a few paces and turned and rode down the line of the various grand boxes of the important personages of France. As he passed his father's box he caught the sight of his parent's proudly nodding from the corner of his eye.

The trumpets blasted their announcement and the heralds called out his name.

The next eight minutes were the most intense that Pierre had ever spent on any horse. The boy put Wingfoot through every conceivable test he could think of. They jumped fences, barrels, ditches and small walls. They cantered and trotted, galloped and danced. Pierre went through an exhausting series of difficult mounts on the move and almost unbelievable trick riding.

At one point, Pierre even leaped Wingfoot over another mount while he was standing erect in his own saddle. This feat drew special praise from the Royal Box. King Charles and his court were truly watching the best horseman in France and, perhaps, in all of Europe.

After his magnificent performance, Pierre galloped back up before the king's box, reined in sharply and then awaited the monarch to speak.

"It is impossible to handle a horse with more

grace or ease" said King Charles with glee. "You are truly an ace. Yes, from now on I will call you Piquez."

Then King Charles turned to the gallery and lifted his arms and shouted out "Piquez, piquez, piquez!". As the king, himself, orchestrated the gallery in shouting the name back three times, Pierre felt himself blushing at so much attention being geared toward him. Thus, though young, Pierre Terrail of Bayard had acquired a nickname of 'ace' and King. Charles had dubbed him the finest horseman in the land.

The games continued and Sir Claude met Lord Bayard's fine knight, Merrick. Each rider splintered his lance on the other's shield on their initial pass. Receiving new lances from their respective squires, the two gallants rode forward, once more. It had been the first time in the tournament that Sir Claude had not unhorsed his opponent with the first try. However, the second siege of Sir Claude proved too much for Sir Merrick and Claude had his fourth victory of the day.

Sir Darwin brought the crowd to its collective feet as he pulled a major upset in the wagering when he unseated the very able Captain d'Ars on the third pass.

The grueling day had taken its toll on the fighters. Now the number of contestants had been drastically reduced to only six. The play was becoming very rough and serious as each contestant was a tough and seasoned warrior who had their own pride on the line as well as the name of their households.

The gallery grew electrified with anticipation as the Black Knight and a giant knight from Spain faced each other. The Spanish knight had a beautiful white stallion which had served him well throughout the contests. With nostrils snorting sweat and ears laid back, the two great steeds mustered their strength and charged along the wooden rails at full speed.
The clash was one of controlled violence as both lances splintered on contact. With a second pass in order, both charged again. Another devastating clash as lances again met shields. Both riders toppled from their mounts and crashed to the ground in an entanglement of flesh, chain mail and heavy armor.
The Black Knight struggled, quickly, to regain his footing. He was a bit groggy but stood, sword in hand, ready to meet whatever aggressive attack that might come.
The valiant Spanish knight made no attack. He made no move, at all, but lay in a still heap where

he had fallen. His faithful white horse stood patiently nearby.

The valiant Spanish challenger would never fight again. His neck had been broken in his fall. The Spaniard, thus, became the fourth fatality of this huge field of the grand tournament.

The saddened gallery hung its head and offered a collective prayer for the fallen warrior.

Pierre watched silently as several men sorrowfully carried the dead Spaniard from the lists and laid him upon his shield. One of the men carrying the shield was the Black Knight, himself.

When things settled back down to the business at hand, the crowd's attention focused on the next battle where Sir Darwin prepared to meet the knight with the Crimson Shield and the blue dagger.

Pierre led the house of Savoy in cheering their champion on. The battle proved to be an exciting one as the two combatants made three passes before Sir Darwin finally sent the cocky young opponent tumbling from his horse.

Following that match, Sir Claude made two passes before unseating a talented Swiss foe and drew great applause as the man slowly limped from the field of combat.

The tremendous field had been reduced to three

gallant fighting men; the best in the land on this particular day.

It would be Sir Darwin pitted against the Black Knight with the winner earning the right to meet Sir Claude de Vaudrey for the championship.

"Would that I were of different station of it" remarked Pierre to no one in particular. " Methinks I would welcome the chance to meet Sir Claude simply for the learning of the experience."

Standing within earshot was Luc LeGroin, who turned to eye Pierre with a wry smile. "Fear not, mon ami" assured the other page. "The rate at which you excel with your weaponry, I say that soon enough that might happen."

"If only it could be?" daydreamed Pierre, lost somewhere in his own thoughts.

Not a single person remained in their seats as all necks strained to get a better view of the action which was about to unfold before them.

Darwin had placed himself in a bit of jeopardy as he had only one lance remaining. This pressured him to have a good first pass or be at a great disadvantage. His big Arabian chestnut colored stallion sent its heels flying against the Black Knight's own battle charger.

Silence filled the air a few seconds before the two lances met as the entire gallery seemed to hold its collective breath.

Coronal met coronal and the two riders passed with neither losing his saddle. The crowd applauded wildly as the contestants turned for another pass. At their master's commands the two great animals snorted and surged forward again.

This time Darwin was struck a stunning blow on his shield. He winced as he felt the shield being torn from his grasp. Sir Darwin righted himself and turned his battle charger to meet his foe for a third time. Both knight's lances were still, miraculously unhurt.

Darwin waited, watching the dark clad figure at the other end of the lists.

Why does he not begin his charge, everyone seemed to be wondering in a single collective thought. The valiant surely could have hit Sir Darwin as the man from Savoy had no shield to ward off any blow he chooses to make. Just as a swelling murmur seemed to be filling the stands, the answer came.

The Black Knight, exercising the epitome of chivalry, graciously was allowing Sir Darwin time to retrieve his lost shield. At a signal from Sir Darwin, his squire, Henri, rushed a new shield to him. Darwin offered a salute to his foe as the gallery vigorously applauded the grand show of sportsmanship.

Again, the two fine knights spurred their mounts forward. The clashes of lances and shields were followed by the grunts of humans taking the punishment of the collision. Both knights fell backward and off their mounts as their armor made a loud crashing sound upon hitting the ground.

Both men were slow getting to their feet. Shaken, they gamely faced each other again. This time it was with swords at the ready. Slowly, too, they circled each other both wary of the other's prowess.

Then, in a sudden movement, they were at each other with broadswords swinging. The ring of steel filled the air as the cheering galleries were privileged to witness some of the finest swordsmanship they had ever seen.

Finally, under one of Darwin's crunching attacks, the Black Knight fell to one knee. His sword point slowly dipped downward until it rested on the ground; signifying that he was yielding to his foe.

Why?

Sir Darwin and the crowd were stunned with a mixture of bewilderment and joy. Why was this magnificent fighter giving up? No killing blow had been landed.

Slowly the man in black removed his helmet and exposed his dark face, which was splattered with crimson. One of Darwin's blows had struck the black helmet with such force that it had shattered the nose of the Black Knight.

The galleries saluted this valiant champion from another land as Sir Darwin removed his own helmet and moved to stand before the man. He helped the black clad figure to his feet and they clasped wrists in recognition of each other's true championship prowess.

"I am injured, Sir Knight." said the Italian. "I could only do myself more harm to continue. Messire Claude needs the best equipped champion to face him if there is hope of defeating him."

"I will do my best." said the man from Savoy. "I shall always remember your skills."

"And I yours." returned the Black Knight.

The crowd again saluted the two gallant warriors. The Black Knight left the field and the heralds announced a brief rest before the final battle of the day.

Sir Darwin would meet the finest swordsman in the land, Claude de Vaudrey.

At Darwin's request, Pierre joined the circle of attendants around Darwin. As Pierre came within earshot of the group he could hear Sir Darwin telling Phillip that his battle charger was too weary to carry him against Claude.

"Sir?" spoke up Pierre, not knowing what possessed him to do so. "I would be honored if you would accept my horse in his stead."

Darwin looked at Pierre as did all others within the circle.

"Fete Dieu." exclaimed Darwin with a broad smile. "That is a kind gesture, young Bayard. But your fine animal deserves a rest after the performance with you. I thank you, though, for your most generous offer. I will use friend Phillip's mount"

Shortly thereafter, Sir Darwin and the champion of all France faced each other for the final contest of the great tournament. To comply with rules set down by the king, himself, Darwin was to meet Claude with sword against lance as Darwin had exhausted his lance arsenal.

This was announced to the crowd, who expressed moans of disappointment as they felt the final battle would be less than perfect with one man being at such an obvious disadvantage. They knew that Claude was an excellent swordsman and probably would beat the knight from Savoy at that challenge. But they were not so sure that Darwin could be taken lance against lance.

Darwin watched the crowd and drew his sword, ready for whatever would come from the champion.

Sir Claude was watching the crowd, too. He was the people's champion and a man who, like the others, thrived on the challenge. With his skill with the sword, Sir Claude felt he needed fear no man. Thus, in a magnanimous gesture, the king's champion passed his lance to his squire and drew his own sword.

The gallery erupted in a great swell of cheering as the two contestants bowed to each other to convey their mutual respect.

As the drums rolled and the trumpets and clarions blared, the two knights bolted to meet each other.

"In honor of Her Grace, Blanche of Savoy." called out Darwin.

The onlookers cheered wildly as the ring of steel filled the air with every blow of sword upon shield and sword upon sword. Darwin was no slouch with a sword in his own right, as Claude soon discovered. During one heated exchange, Darwin sent his foe's shield flying from his grasp.

Darwin turned his charger and discovered the king's champion with no shield. Savoy's champion saluted Claude and promptly returned the favor given to him earlier by the king's man. Darwin hurled his own shield to the ground and dismounted, to the pure delight of the gallery.

Claude returned Darwin's salute and quickly dismounted himself.

For the next quarter of an hour the host of

onlookers were provided with some of the finest swordsmanship they or anyone had ever witnessed. When both fighters were so arm weary that they could scarcely lift their great heavy blades, the young king put the decision directly to the contestants, themselves.

The two knights had gained a tremendous amount of respect for each other.

"We could stay at this till nightfall, methinks." said Darwin. "You are the best I have ever seen."

"You, sir knight, are the finest I have ever faced." retorted Claude. "What say you about the outcome?"

"Of a truth, Sir Claude." answered Darwin. "Over a given time, perhaps you would beat me down. But I have a strong arm and youth on my side and the sun is getting very hot."

"Well, there is no life and death matter at stake." agreed Claude. "I must be getting old as I have desire only to live out my days in peace. What say you, we call this event even and face each other, perhaps, next year?"

"You are still the champion of all of France, no matter the outcome." assured Darwin.

"You are quite the champion, yourself, Darwin of Savoy." complimented Claude. "I will drink to it at the banquet this night."

Thus the two great knights decided the end of the

perfect tournament by mutual consent and King Charles eagerly led the applause.

Then the heralds cried "Fold Your Banners" and the great tournament was ended. Soon the lists and galleries were emptied as everyone went to bath and dress for the grand finally of the tournament, the king's majestic banquet.

For hours, feasting and dancing and storytelling and joking filled the great hall. Then the newly crowned 'Queen of Love and Beauty' was seated and each proud knight who had won a prize was led before her. To each one, she offered a well rehearsed speech which always concluded with the hope that the particular recipient would be 'happy with his lady love'.

Each knight was most careful to make the proper response. "The victory was owing to the favor of my lady, which I wore upon my sleeve, my lady." was their answer.

After the prizes were distributed, gifts were given to the heralds for their hard and diligent work throughout the tournament.

The final part of the king's great banquet then took place after Pierre proudly accepted a silken tunic of silver embroidery for his prize.

Then it was time to award the final prize. This

prize did not go to the man of the most noble birth but, rather, to the contestant who exemplified the highest form of valor and chivalry of the tournament. Before this year, that coveted prize was never shared by three men.

Sir Darwin of Savoy, Claude de Vaudrey and the Black Knight from Italy happily shared in the honor.

King Charles then asked Anne, as 'Queen of Love and Beauty' to honor him with a dance. Then Claude, the Black Knight and Darwin begged their turns and finally Pierre, the Page, claimed the last dance of the night. He was the proudest person in the entire room.

CHAPTER FIVE

Even though the tournament had been a great success and the House of Savoy admirably distinguished itself, Duke Charles was not making good headway with his exalted cousin. The king had been only barely cordial and polite to the young duke, who became offended and took it as a personal affront that, not once during this stay, did the invitation come from the Royal Court to set at the king's own table. Thus, it was with battered pride that the embittered young Duke of Savoy sent half of his entourage back home and continued with the remaining half to visit the cities of Chateau Renault and Vendome.

Pierre was among those sent back home to continue his training.

"Methinks our days of it might just have gotten more complicated." pointed out Pierre's chief rival in all things of training thus far. Luc Le Groing began to tell Pierre of Duke Charles' return from Vendome, which was to be within the hour according to a messenger sent on ahead of the party. The Duke and Duchess also had, again, toured in Lyons and had brought back four new people who would be permanent members of the household.

Le Groing related how one of these was a boy barely a year older than Pierre. He was related to a nobleman who was with the House of d'Ligny. Pierre, of course, was very familiar with the name of Paul of Luxembourg,

The Count of d'Ligny, the future Prince of Altamura and of Venosa. The Count was in his early twenties and handsome of face and reputed to be very generous and brave. He was one of the brilliant new generation who had more assurance than knowledge and the likes of whom was systematically replacing members of the French cabinet that were hand picked favorites of the former King. A first cousin of the King, himself, the Count stood high in the esteem of the Royal Court. Young Paul of Luxembourg also was the proud owner of one of the finest Schools of the training at arms of any in the whole of France.

Curiously, Pierre prodded into his friend's meaning of how life would become more complicated with the addition of the newcomers.

"What means you, friend Le Groing? Of life getting tougher? "With a small smile, the lad from Bayard added "methinks they are right tough enough now."

"Rumor is that accompanying His Grace is a new Page of right good repute with both horse and arms." answered Luc Le Groing. "If rumors are true, we both might well be put to the new test sooner than later, methinks."

"Rumors abound, friend Luc" said Pierre with a small laugh. "However, if they be true and we do have new blood then more the good of it, methinks, because it will only bring out the best in us."

"As usual, you are right, mon ami" chuckled the other page. "At any rate, we should know of it within the hour as they will surely arrive by then."

Pierre de Poquieres, de Bellabre et de la Marche. Certainly, that was quite a mouthful for any man's name, much less for a young man of only sixteen.

Pierre had only recently been appointed the new Chief Page. He knew that any newcomer of the rank would have to be brought to him for induction and equipment. So, whatever the cards of fate would hold for himself and the newcomer, Pierre knew it would reveal itself shortly.

Pierre looked up as the big door swing open and he could see, Luc, himself, coming his way with the one that had to be the newcomer beside him. The young man was very close to his own size in height and weight and general appearance. In fact, had one not known better, the two boys might easily be mistaken for family. Like Pierre, the newcomer had a pleasant smile and alert, grey eyes.

"This is Pierre de Poquieres...uh." began Le Groing as he was obviously having some difficulty pronouncing the name.

"Pierre de Poquieres, de Bellabre et de la Marche." introduced the young newcomer boldly, with a smile.

"Easy enough for you to say." chided Pierre with a good natured smile. "I am Pierre Bayard."

"You are the one." remarked the newcomer as he studied the boy from Bayard closely. "Please, just address me as Pierre Bellabre." offered the new Page. "I am pleased to meet you as your name is famous in our house."

"You mean my family name." corrected Pierre with his usual modesty.

"Oh, no, I mean your name." assured Bellabre. "For you see, you seem to be the measuring rod that each page must attempt to live up too. Believe me, it is no easy chore of it" said Bellabre with a small laugh. "So I am pleased to meet you, friend Bayard. Oh, I guess I already said that, did I not?"

Pierre smiled but said nothing as he recalled his own nervousness on his first day at Savoy.

"I look forward to witnessing your riding." said Bellabre. "For I sit a fair Palfrey myself, as well as schooled in my arms and I am told and they say that any man can learn much from the doing of it. I do so wish to give a good account of myself"

"As it should be, friend Bellabre." said Pierre matter-of-factly.

"We shall have to watch this one, friend Bayard." said Le Groing with a friendly smile.

"We all wish that." said Pierre.

"I am fairly well trained at my skills." said Bellabre. "I am looking forward to pitting them against you, Pierre Bayard."

"I just train as I train." said Pierre. "But if you are as good as I am hearing that you are, then it will be good to have a fresh challenge."

"Oh, you two are going to be a lot of fun to be around." called out Le Groing as he turned to leave. "This should be interesting, indeed."

The two Pierre's watched Luc Le Groing walk away. "He seems to be a proper one." observed the newcomer. "I suspect you two have had a go or two at each other."

"Indeed, friend Le Groing is a tough competitor." admitted Pierre. "What he lacks in size he makes up for in skill and heart."

"And stubbornness, methinks." said Bellabre with a smile.

"And stubbornness." agreed Pierre. "Come friend, Bellabre. Let me show you around and get you properly housed and clothed."

The two Pierres soon discovered that they had their heated moments in rivalry, as was expected and Bellabre held his own on many accounts. However, they discovered something else, as well.

They were quickly becoming the very best of friends. Both were strong, quick of wit, and full of talent. Pierre had finally met someone with whom he could share all things and someone who was completely open as a friend.

The two Pierres seemed to share just about everything with each other including their taste for beautiful women. However, there is where Bellabre seemed to be wise enough to draw the line as he recognized that Anne belonged to Pierre alone. The two soon became inseparable as they formed a bond as close as any two brothers could have done. In effect, they became brothers.

While the two Pierre's friendship became cemented for what would be life, things were not going so well with young Duke Charles and the household of Savoy.

The duke's feelings had slowly turned against his Royal cousin after the cool treatment he had received at the Royal Court. Little by little, the head-strong duke began to withdraw his household and its support from the side of the French and align it, more and more, with his Italian confederates.

This greatly began to disturb the likes of Bayard and Bellabre and Le Groing, who were all true Frenchmen from solid French families. Yet, at the same time, they really liked the energetic young duke and duchess and were honor-bound to serve them to the best of their abilities, within reason.

Just when the question of the status of Duke Charles' loyalty to France was beginning to become a gigantic concern for the three youths, a tragic and unexpected turn of events occurred early in March of 1490.

Young Duke Charles of Savoy became deathly ill with consumption, albeit the rumors flew that he had been poisoned and was confined to his bed. He spent ten pain-wracked days under close supervision of his physicians, whom everyone said were incompetent to begin with.

Thus, a great sadness engulfed the House of Savoy on March 14, 1490 as Duke Charles, while still in his early twenties, died in his sleep.

For five rain-filled days Charles 'The Warrior' lay in State at Pinerolo as his loyal Pages and Squires took turns around the clock standing guard. Even though some claimed that his surname 'The Warrior' was, perhaps, a bit too easily acquired via his small campaigns against Saluzzo, he was none-the-less loved and would be greatly missed.

The following six months were not good ones for Savoy. The young and inexperienced Duchess Blanche found that her husband had left a tremendous debt of 300,000 Florins. The situation began to look almost hopeless for Blanche. Finally, she had to call a house meeting and everyone expected the bad news even before she gave it. She simply had no choice but to let about one half of her household staff go. She had to lessen her burden or lose everything she had.

The various servants were set free to find employment elsewhere and Blanche even helped some of her favorites by getting them placed in other good houses.

Only four Pages and eighteen Squires were to remain with the eighteen knights that Blanche could afford to keep until she could make other arrangements. The Two Pierres along with Lucquin Le Groing and Jacques de Montbel de Varel were the pages that were to remain. Rumors abounded as to the fate of the young duchess, who obviously was facing much more than she could cope with.

The winds of hope blew around the thought that, somehow Savoy would survive.

However, the situation at Savoy only worsened.

The very instant that Pierre caught sight of her, almost running toward him, he bristled at whatever it was that had upset his lady love so much.

Before he could even ask any questions, Anne was holding her body tightly against his and sobbing uncontrollably.

"Pray tell me. What is the cause of your misery?" he anxiously asked quietly.

Anne looked up at him and into his eyes for the first time.

"Pierre, my love, they are sending me away." sobbed the girl. "I will never see you again. They can't do this."

Through Anne's sobbing, Pierre quickly learned that one of the moments that he dreaded so much was finally at hand. It only stood to reason that Blanche would have to finally come to this stage in the ugly game of court politics. Little by little the good Lady Blanche was being forced, out of necessity, to sell everything simply to stay out of debt. It had begun! Anne being sent home was just another step in the long and sad demise of the proud House of Savoy.

"We will meet again." vowed the heartbroken young Page with as much courage as he could manage to muster under the devastating circumstances.

Pierre found that saying goodbye to Anne was even tougher that it had been to say his goodbyes to his family before he was sent off to Savoy.

Blanche finally announced that she had to sell off the House of Savoy and all of the lands and holdings. She took special pride in her young pages and made special arrangements for them. Duke Charles' uncle was the powerful Lord Phillipe de Bresse who was only too happy to receive the quartet of fine young pages as he, himself, was destined to become the next Duke of Savoy within six years' time.

Thus, Pierre Bayard, at sixteen years of age, left Turin early in the afternoon on October 5th, 1490 and was headed for Grenoble. Pierre Bellabre, Lucquin Le Groing and Varel would be following in a week or two.

Pierre arrived in Grenoble only a day or two ahead of Phillipe de Bresse and got to spend some time in a very pleasant visit with his family. Pierre, however, grew concerned over his father, who was much sicker than he would admit.

Pierre's future seemed a little bit topsy-turvy as Phillipe de Bresse faired only slightly better than his nephew before him had with the impetuous Royal Court. After only a few months, Phillipe refused to be caught as a pawn in a political battle and decided to cut his losses and retire in comfort while he had the chance.

The Pages, also unwilling to be left with nothing nor no place to go, held a meeting of their own. It was Jacques de Montbel de Varel, who came to the four boy's rescue. His older brother was the personal secretary to a very distinguished nobleman of Lyons, none other than Paul of Luxembourg, the Count d'Ligny.

Of course, Pierre had heard of the dashing young Count who was but in his twenties himself but who was destined to become the future Prince of

Altamura, a title which carried with it the Duke of Andrea and Venosa, Lord of Vohhera and of Tortona and the governorship of Picardy.

He also just happened to be the King's own First cousin and was one of King Charles' most capable military advisors and a battle-hardened warrior.

His own father, the Constable of Saint-Pol, was put to death at the hands of the former King Louis XI. Despite this small handicap, d'Ligny was held in high trust in the new French Court. He had even succeeded in regaining all of his families lost estates and now was even the proud commander of a hundred lances in King Charles' own army; a privilege accorded only a special few and most of those of Royal blood.

The Count was a brave and popular young lord, indeed, and was impetuous, head-strong and temperamental. Yet, he was very good at what he did and seemed to do all the right things to keep the Royal Court well pleased.

However, perhaps even more importantly, to Pierre and the other Pages, the good Count owned the best military training academy in all of France.

It was a stroke of good fortune and a very proud day in the lives of the four young Frenchmen when confirmation came of their acceptance into d'Ligny's service. With this great leader, Pierre's intimate education in French military service began under far more varied and dynamic circumstances than his young life had known to this point. The training became even more grueling and demanding under the watchful eye of a gallant Captain, d'Ligny's chief lieutenant, none other than Captain Louis d'Ars. Pierre could not think of a better setting to be thrust into than to be trained and guided by the man whom he had chosen for his personal hero some time back.

There was a never ending stream of information to be absorbed. Pierre and the others learned to condition their bodies to meet all kinds of hazards. They learned to climb jagged cliffs, run faster and farther than ever before. With all that was transpiring in his busy young life, Pierre, although thoughts of her came to him now and again, found himself thinking less and less of Anne and more and more of what his future might hold.

There it was. Pierre had been standing quietly beside Wingfoot and simply looking, almost in awe, of it for a good five minutes now. The other Pages and Squires were all doing the same thing.

The day had finally come, much to some of their chagrin. This was the first time that any of the new pages had even seen the Quintain, let alone would be facing it.

Facing the dreaded Quintain was, perhaps, the single best experience of a young page or squire's preparation for hand to hand combat from the back of his moving mount.

It did not take the two Pierre's long to figure out just how this man-made monster worked. It was simple enough looking; a post erected with a simple crosspiece of wood that would whirl at the slightest touch of a lance.

From one end of this crosspiece hung a board and from the other end hung a heavy sandbag.

The rules, too, were simple. All one had to do was to ride at the Quintain at full speed and strike the board with his lance.

"However." warned d'Ligny, who was this day personally taking on the training of his dozen new pages and squires as he often enjoyed doing. "Do not let its look of innocence betray you, gentlemen. Ready yourselves for you will be in for a battle, I promise you."

Pages and squires, to the man, let their gazes scrutinize the Quintain before them, which was garbed loosely to resemble a knight. It was difficult for them to imagine something as simple as this as having the capability of hurting them. In fact, probably everyone there was telling himself that he would be the exception to the rule and would beat this man-trap.

"Yes, I must admit that it looks rather easy." said the Count, almost as if he were reading their minds as he gave Captain d'Ars a wry knowing smile. "But, my advice to you would be to drive that thought far from your heads. It is NOT easy. Tell them, good Captain."

"Of a truth." spoke up the youthful but much celebrated Captain d'Ars. "If you become the least bit slow or clumsy in your run at Sir Quintain, you will be the worse off for it. You will pay the price. The Quintain is very quick, like the very strike of a snake and every bit as harmful to you."

Even with this further warning, the candidates mumbled, disbelievingly, among themselves. All except the two Pierres and the tough Luc Le Groing. Somehow, they knew what might happen. Yet, even they proposed that if a rider but would keep a cool head, he could fair well against this thing.

"Gentlemen!" called out Count d'Ligny as the usual crowd of onlookers began to gather as they always did at the battle with the Quintain. "I issue this final word to you. The Quintain never misses his stroke so take care that you do not miss your own."

With these unpleasant thoughts, the young pupils

took their places in line as the Count, Captain d'Ars and a sizeable gathering of onlookers seeking amusement waited.

The first to test his metal against the Quintain was one, Gaspard de Coligny who was a young noble with better than average skill. He surged his mount forward with full confidence.

Coligny's lance struck the Quintain's shield a stout blow and, immediately, the sandbag whirled and found its waiting target on the back of the young man. Coligny was sent sprawling, head over heels, to the ground amid joyous laughter from the spectators and smiles from the trainers.

The Count motioned for the next trainee in line to have his go at Sir Quintain. That was Luc Le Groing.

As had Coligny's lance had done, Luc's lance hit the shield square in the center. As did Coligny, Le Groing was toppled from his saddle amid the good natured laughing and cat calls from the onlookers.

Three squires followed and met the exact same fate as had the two gifted pages before them.

Then all eyes centered on the young man from Bayard as Pierre dropped his lance and prodded Wingfoot forward. Bending low in his saddle Pierre watched the distance between himself and the Quintain close quickly. Pierre's lance struck a true blow on the wooden shield. The Quintain began to spin and the sandbag came around like a shot from a cannon. Pierre could feel the wind of the sandbag as it passed within an inch of his head and he let out a great sigh. Then he heard loud cheering and calls of "Bon!" and "Bravo" and "Well done."

"Well done, young Bayard." said Count d'Ligny, himself, as he eyed Pierre. "Now do it again and we will see if it was luck or skill that guided you."

Again the young man from Bayard rode toward the Quintain. Again his lance struck the wooden shield. This time the sandbag whirled so quickly that it caught his left shoulder a glancing blow. Pierre reeled in his saddle as the onlookers let out a collective gasp; knowing that he was going to fall just like the others had. But Pierre wasn't like the others. He bent backward, almost having his head touch the ground, then, with strong legs, regained his mounted position to the amazement of all watching.

Again, the cheering was tremendous after a brief, collective disbelieving gasp sprang from the crowd of onlookers.

Count d'Ligny and Captain d'Ars exchanged quick silent glances then wheeled their mounts and rode up to flank Pierre on either side.

"I darest confess that it has been a very long time

since I have witnessed anyone making two straight and successful passes at Sir Quintain" said d'Ligny.

"What of you, my lord, Captain?"

Captain d'Ars shook his head in near disbelief. "Methinks that this young sir is, indeed, the horseman that all have claimed him to be."

"Of a truth, you are right." agreed d'Ligny with a broad smile and a slight nod toward the newcomer from Castle Bayard.

With that, the two soldiers wheeled their horses and rode away from the boy from Bayard as Count d'Ligny yelled out "Next!"

Claude Jacques de Miolans was next and promptly found himself on the ground to the tune of merry laughter from the onlookers.

One after the other, throughout the next full hour, the challengers continued to ride against the Quintain. Time and again, one tough squire or page pitted their skills against those of the whirling monster. One by one, the likes of Guillaume d'Hermance, Jacques de Montbel de Varel, Claude Charles de Montagny and John Erlach of Bern all attacked the Quintain and met the same failing humility.

The final contestant of the day was Pierre Bellabre. With a heart that seemed to be in his throat and great anxiety, Bellabre took a deep breath and charged. He struck the Quintain as he flew by so fast and ducked so low that the sandbag 'swooshed' harmlessly over his head.

"Nothing too it" bubbled Bellabre in happy pomp as he rode up to Pierre.

"Glad you feel that way" laughed Pierre, "because the Count is motioning for you to do it again."

Once more, Bellabre's lance struck the heart of the wooden shield. Once more the sandbag spun. This time it found its mark on Bellabre's shoulder. The rider reeled, awkwardly, in his saddle, rode a few yards before half falling and half leaping from his horse and hit the ground running. Only a collision with Wingfoot kept the young man from toppling over a railing.

From atop his mount, Pierre looked amusingly down at Bellabre. "Nothing to it that time, either, I guess. Listen to how they applaud you." he said.

With that, Pierre wheeled Wingfoot away. The suddenness of the horse turning left Bellabre leaning on air. He fell to the ground amid more laughter.

"A fine way to treat a friend." called out Bellabre playfully to Pierre.

Each day, for weeks, the contestants pitted their skills against the Quintain. Little by little they grew better and more adept at facing the monster. Only once did Pierre find himself swatted from the saddle. That feeling of hitting the hard ground was one that he did not wish to experience often.

As the weeks turned into tough months, the training grew even more intense for the young candidates. Count d'Ligny had one of the top schools in the land. Of all the worthy candidates in the count's service, Pierre Bayard and his inseparable companion, Pierre Bellabre proved to stand out at the top of almost every test they were confronted with. Count d'Ligny and Captain d'Ars were well pleased.

CHAPTER SIX

Pierre received a very welcomed early birthday present just before his seventeenth birthday.

Paul of Luxembourg announced the results of his semi-annual in rank promotions for Castle d'Ligny. Two received their long awaited commissions into the 'brotherhood of lances' as they were given their golden spurs; having become full-fledged knights.

Four pages were extremely happy young men when the announcement was made that they had graduated to the prestigious position of 'Gentlemen-in-waiting'. Pierre and his constant companion Pierre Bellabre along with the tough Lucquin Le Groing and Gaspard De Coligny were all named to be the proud new Squires of the d'Ligny household.

Just two days later, the House of d'Ligny was honored with a very prestigious guest in none other than Claude de Vaudrey, himself. Of course, the news had everyone buzzing with excitement. The great champion of France had come, at the request of the King, himself. It appears that the good Chevalier, not one to like the long days of idleness between action, had convinced His Majesty that to keep the soldiery on their toes, that he would offer himself as a training tool for the best riders in the best school in the land. The king readily agreed and sent Sir Claude straight to Burgundy and the Count's excellent school.

The announcement had been properly posted that Sir Claude would align his shield so that three days from then any who dared could strike them wherein they would soon meet the great champion both on horse and on foot for the thrusts of lance and blows of ax.

Pierre stood, almost mesmerized by the sight of them.

"A grand and formidable sight methinks" said Bellabre quietly as he stood beside his best friend. "Only three have taken up the challenge so far" remarked Bellabre. "And methinks not many others will even bother to try."

"Fete Diu" said Pierre quietly. "If I but knew of how to equip myself, how willingly I would touch the shields."

"Of a truth" exclaimed Bellabre as he gazed unbelievingly at his friend.

"Has the sun touched you too much or have you simply gone mad without any help?"

"I mean it, mon ami." said Pierre. "Oh, I know what you are thinking and others might think. But, of a truth, it is not of pride that I seek such but for the learning of the thing alone."

"But you have been a Squire for only three days," protested Bellabre.

"I care not a fig of that" assured Pierre in all seriousness. "By his kindness, my Lord d'Ligny has seen fit to raise me from order of a Page and placed me amongst the order of gentlemen. It seems to me that it would be of great honor to him for me to touch these shields. But, alas, I am doomed of it from the start as I know not who would furnish me with the equipment and the horses. For my purse is empty."

There was a long silence as each squire stood quietly in his own thoughts.

Bellabre had proven to be every bit the adventurer and fun loving person that Pierre was and had grown to love his new best friend more than life itself.

It was Bellabre who broke the silence.

"By all the saints" he said, almost shouting the words, "If you are really serious then methinks there is a way of it."

"How, I haven't two francs to rub together."

"Of a truth, you are right, mon amis" agreed Bellabre with a sly smile on his face.

Pierre could scarcely miss the look on his friend's face and knew Bellabre well enough by now to know that he was up to something. Before he could ask, Bellabre was talking excitedly again. "Did you not say your kinsman is the Abbot of Ainay?"

"To be sure, it is the truth." admitted Pierre. "But what does my uncle have to do with it?"

"Everything, dullard, everything." chided Bellabre as he stepped back and grinned broadly.

"I don't know." said Pierre with a small shake of his head.

"What is this?" mused Bellabre. "Are you afraid of your own uncle? My wager of it is that once he hears of your desire, he will willingly help you."

Seeing Pierre's reluctance at the proposal, Bellabre prodded. "You desire to ride against Sir Claude, do you not?"

"On my faith, that is all I seem to want at this moment." admitted Pierre.

"You further do not have armor. Is that not correct?" further baited Pierre's friend. "And is it not true that you do not have coin to get what you seek?"

"Well... yes... I "admitted Pierre.

"Well, Sir Squire. Neither have I the coins or they would be yours" said Bellabre. "So, you must decide."

Pierre well understood that Bellabre was right. Asking his famous uncle for help seemed to be the only way to make his dream of riding against Sir

Claude happen.

With a confirming nod that sealed his decision, Pierre quickly mounted Wingfoot and guided the great horse in front of all five of Sir Claude's shields and boldly struck each one, to which even Bellabre stood wide eyed in disbelief.

Seated nearby was the Royal King-at-arms, Montjoi, who took on a look of pure amusement at the new squire's bold actions.

"Hallo, friend Squire." called out Montjoi. "You have not had even a beginning of a beard and you dare undertake a fight with one of our countries most accomplished gentlemen?"

Pierre reigned in Wingfoot before the Court Recorder. "Friend, Montjoi." he said with a smile. "That which I do is not out of pride nor overconfidence. Is it against the rules?"

"No." admitted the still puzzled king-at-arms.

"Then, I, Pierre Bayard, wish the challenge to stand as my only motive is curiosity of what it would be like to face one of the best in the land." answered Pierre with unabashed clarity. "Seeing that blood of battle is not involved and if little by little from whomever I may learn that I become a strong arm for my Lord d'Ligny and if God, in His good grace can allow me to do that which pleaseth the fair ladies, then methinks no harm can come of the doing of it."

Montjoi let out a hearty guttural laugh, shook his head in disbelief and waved Pierre on. "Friend Squire, if it can be done then, on my oath, I do believe that you would be the one to accomplish such a monumental task."

As Pierre rejoined Bellabre, who had already mounted his own horse, Pierre quietly confessed "Would that I have not taken on more than I bargained for." Then he looked at his friend. "I must seek permission to have audience with my kinfolk. Are you coming?"

"Oh, I would not miss this for a month's pay." assured Bellabre as he spurred his horse to keep up with Pierre, who was already rounding the corner of the building in front of them.

The Abby of Ainay was on a small island some six hundred yards into the middle of the lake. After gaining permission to visit his uncle, the two Pierres secured the services and paid a boatman what Bellabre claimed was his last two farthings. Probably even before the two squires reached the Abby, word had spread like wildfire of Pierre's touching all the shields for Sir Claude. Some praised the young squire for his bravery and some dubbed him a fool for the doing of it. Whatever the talk, it grew until it finally reached the ears of Count d'Ligny and Captain d'Ars.

Nearby was the King, himself, with some important guests including Claude de Vaudrey. Seeing his exalted cousin and Captain d'Ars in what seemed like a hushed conversation, King Charles called out "Good cousin, d'Ligny. Is there trouble brewing?"

"No Sire!" called back a bemused d'Ligny as he shared a knowing smile with his favorite Captain. Then the two soldiers strolled over to the monarch and his guests. "Well, no trouble for the crown, that is. But methinks one of my squires doth take on much more than he can handle in a rather foolish deed. Yet, my good Captain d'Ars, who is the lad's chief trainer, seems to think otherwise."

King Charles smiles "Of a truth, cousin, what has the squire done that has your attention, so?"

"Your Majesty will find pleasantry in this" began d'Ligny, "as it is your favorite young Bayard."

"Piques? What has he done, pray tell?"

"Well Sire" answered the count "it seems that Bayard, only three days a squire, has made a bold but, methinks, a rather foolish move." Then the Count, with a smile, looked at Sir Claude. "It does include you, my Lord de Vaudrey."

Sir Claude simply raised his eyebrows but said nothing as he did not take into account anything that a mere squire might have to do with him.

"It seems young Bayard has boldly touched all of

your shields." announced the Count.

"A squire?" questioned the visitor.

"Upon my soul, you know of him, friend de Vaudrey" said the king in a joyous voice. "For it is the same lad who did ride before me at the grand tournament."

"The boy horseman?" quizzed the knight.

"Indeed, the same." answered King Charles, with a chuckle. "Only now he is of seventeen or eighteen and quite accomplished with his arms."

"You did open the invitation to all." reminded Count d'Ligny with a small chuckle. "However, perhaps it is too brash a move for a Squire to take. Methinks he is still too young to take the thrusts of one of your caliber. I will quit him of it if you desire."

"Oh, by my soul, cousin! Please do not do that." begged the king. "Methinks it might be a boon to the moral of the gentlemen of our companies who do grow more restless daily for action. I deem it marvelous, indeed, that one so young does show such courage. Perhaps it will spur some others to take closer look at their own training for to see what might come of it."

"If Your Majesty wishes it and Sir Claude does consent of it."

"I have no quarrel with it." assured the seasoned knight. "It might add an amusing page to the story."

"The matter is settled, then." conceded d'Ligny.

"Oh, of a truth, I am pleased" bubbled the King sounding like an excited school boy. "I have confidence that your young charge will present himself quite well."

Meanwhile, the two Pierres stepped from the boat and went to find the Abbot after instructing the boatman to wait for them.

They came upon a monk who was sweeping a sidewalk and they inquired as to where the Abbot could be found. The monk told them the Abbot and a companion were in the garden.

As the visitors were approaching along the garden path, they could see the Abbot and a monk sitting on a bench. It was obvious from their actions that the two men of God were surprised to see anyone in the garden with them. The Abbot was just finishing taking a swig of something from a small container. By the looks on the Abbot and his companions' faces, it was also obvious that they were caught drinking. As much as they attempted to hide their actions from the two visitors, they had not gone undetected by the sharp eyes of the two Squires.

The monk held up a staying hand to the newcomers. But, the Abbot, upon recognizing his nephew, waved them to him.

"How very much you have grown to resemble your Grandfather, our kinsman." said the Abbot before giving Bellabre a quick glance and nod of recognition. "What brings you before me?"

"Uncle, most learned and loved man of our family" began Pierre, being a bit inwardly surprised at his own nervousness. After all, this was his own uncle.

"I now serve the House of d'Ligny as gentleman in waiting and I... well... I..."

Bellabre, upon seeing his friend's difficulty and in a way only he could have gotten away with, quickly jumped into the conversation.

"What my friend is trying to say, your Grace" explained Bellabre. "If I may speak for one who seems to be far too modest and shy as to talk for himself and seeing that you come from the same honorable family and seeing that he is of a mere seventeen summers, wishes to let you know that he has a deep desire to follow in the footsteps of those warriors who have come before him." Pausing just long enough to see that the Abbot was listening, Bellabre boldly continued. "Therein, knowing of the fine reputation that is held of being you and your Abby, he wishes nothing to impair the same."

"Hold the chatter, young sir." interrupted the Abbot with a slight frown before turning his full attention to his young nephew. "What is the talk I hear of some foolishness that has come upon you?"

The words took Pierre by complete surprise. This was not the response he had been hoping for from his uncle. It was pretty obvious that the man had heard of what Pierre was planning to do and made no pretense of hiding his feelings. "Well, master boastful" continued the Abbot, "from where comes your temerity?"

"I assure you, my uncle" stammered Pierre, "it is not pride that I do what I do. My desire and wish is only to come by virtuous deeds, to the same honor of my forefathers – and your own kinsmen – have done on the field of battle for France."

This time it was Pierre who gave his uncle no time to respond as he was already talking again. "Therefore I humbly entreat of you, my lord, as much as I am able and seeing that I have naught friend nor family here, except yourself, that it be your good pleasure to grant me aid of the doing of that which is needed to obtain that which is necessary for me."

"My pleasure?" answered the Abbot in more than a small growl in his voice. "By my faith, boy, you can well seek elsewhere of your desires. The people did not give their coin to the church for you to squander on folly. Ney, I say."

Pierre felt deeply rejected and began to resign himself to the fact that his uncle was not going to help him fulfill his greatest desires of the moment.

Bellabre, every bit as keen as Pierre in his own right and seeing of his friend's hurt, spoke quietly to his friend. "Let me reason with your uncle and methinks he will consent to do that which you wish."

Before Pierre could answer, the very set minded Bellabre moved back to the Abbot and firmly took him by his arm and lead him three or four steps away against mild protest from the man of the church.

"Friend, Montseigneur" spoke out the Squire "You and I both understand that it is only because of the influence of your exalted and much loved family that you do hold your position of this place."

Even though Pierre's uncle bristled slightly at the words, Bellabre gave him no time to verbally respond as he was, again, speaking. "Therefore, methinks a wise man such as yourself would reflect upon such truth and be willing to return such favor to a kinsman such as my companion, your nephew, whom the king loves greatly. Methinks it would be right prudent of you to grant him his wishes for the good of it will surely reach His Majesties ears and he will be well pleased of it and is honor bound to compensate you many times over for your kindness."

It was obvious that the Abbot was studying

Bellabre's words and might have made a decision all by himself. However, Bellabre was not going to let such an opportunity pass without playing what he knew would be his trump card.

In even a quieter voice, the Squire leaned closer to the Abbot and all but whispered. "Of a truth, that which was witnessed upon the arrival of my friend and myself need not be known beyond this garden as I wager your position at this place would become as a bobble dangling from a thread if certain people were to hear of it."

This, immediately, captured the Abbot's full attention as he gulped a small gulp of uncomfortableness. He could not afford his followers to find that he was a drinking man. It could lead to consequences that he neither needed nor wanted.

In a complete reversal of attitudes, the Abbot looked at Bellabre and asked "Of a truth, Sir Squire. What sum are we speaking of?"

Inwardly smiling at his apparent victory, Bellabre studied for a long moment, mostly simply to keep the churchman squirming uncomfortably as he knew, well, the amount. "The sum, whatever given above a hundred crowns would surely be conveyed to His Majesty, who would think highly of one who brought aid to his favorite squire." As Bellabre gave his answer, he watched the Abbot's face intently and could see the man already contemplating his reward.

Then the Abbot, as if simply wanting to close this matter before this upstart young 'blackmailer' could have something more to hold over his head, nodded slightly and without another word, motioned for the two visitors to accompany him.

Pierre's uncle took them straight to his quarters where he went to a cupboard and withdrew a small box. From the box, he took some coins and placed them in Bellabre's hand. "This will buy horses that are needed in this venture," said the Abbot. "I give it unto your keeping as my nephew is still to young in years to act for himself in such matters."

"We thank you, Montseigneur" said Bellabre with a slight bow. "And of the armor?"

The Abbot shot Pierre a quick look, which made the mildly confused young man from Bayard a little nervous, as he still didn't quite understand all that was taking place before him. "I will write a line unto Laurencin to furnish him with what you require."

"It is a deed well done, my Lord Abbot" assured Bellabre as he dropped the coins into his purse. "And methinks when all men know of it, you will be much praised and overly compensated for the doing of it."

Bewildered at his uncle's sudden change of

attitude but, none the less very grateful and pleased, Pierre watched the Abbot write upon a piece of paper and hand it to Bellabre. "This will suffice for my nephew to acquire the hundred francs or so of merchandise that he is in need of." said the Abbot as he shot Pierre a less than friendly look. "I still think it is folly for him, a mere squire, to dream of crossing blades with one so accomplished."

Both squires thanked the Abbot most heartily for what he was doing then left to return to the boat before Pierre's uncle could withdraw his gift.

Once in the boat and on their way back to shore, Pierre turned to his friend with a great puzzled look. "It puzzles me, friend, Bellabre, of how my uncle did change his mind so quickly on the matter put before him. What was it that you told him?" asked Pierre.

"Of a truth" assured Bellabre with a small chuckle. "I simply presented the Lord Abbot with some facts, the likes of which he could not afford to ignore."

"Well, whatever it was, it accomplished our mission." admitted Pierre.

"Yes it did, mon ami" said Bellabre with a smile. "It did that."

Then Bellabre sat down and opened the paper which he had received from the Abbot to make sure that the words would not, somehow, vanish from off the paper he held.

Suddenly Bellabre's face broke into a huge smile and he gave a joyous shout, which commanded both Pierre's and the boatman's full attention.

"Methinks God has, indeed, seen fit to equip you in any manner you choose, mon ami" stated an excited Bellabre. "You must take full advantage of it."

Pierre simply shook his head as he was attempting to make some sense out of his friend's words. What possibly could have Bellabre so excited, wondered the boy from Bayard.

"Look. Look at this." said the squire as he waved the paper that the Abbot had given him in front of Pierre's face. "Dame Fortune has, indeed, turned her face upon us… and you."

Pierre eyed his friend warily as he was unsure of what Bellabre was so excited about but was pretty sure that he, somehow, would not like it as much as Bellabre would have it.

"What are you talking about?"

"Your Uncle has blundered, methinks, and to his remorse when he does discover what he has done."

Pierre gave Bellabre a puzzled look and was getting an uneasy feeling at what he might be hearing next.

"Here. Your esteemed Uncle seems to have

forgotten to put a limit on his generosity. He has signed a paper that offers you what you want for the mere taking of it" bubbled Bellabre with great enthusiasm.

Pierre sat silently looking at, but not really seeing, the paper that his friend was waving in front of him. The boy from Bayard had an uneasy feeling about what he knew Bellabre was proposing. Somehow it simply didn't seem right to trick a holy man.

"Oh, I don't know...."

"Come now, friend Pierre." interrupted Bellabre with great conviction and seeing Pierre balk at what was being laid out before him.

"We must take full advantage of the opportunity that God has given to us. Did you not tell me that the Abbot was your favorite uncle?"

Pierre was going to answer but Bellabre didn't give him a chance as he excitedly continued. "And do you not suppose that you just could be his favorite nephew? As such, can you not see where this is his way of making sure that you succeed in your quest?"

"Well... I...."

Again the excited Bellabre cut him off and was laying out a very plausible reasoning as he had a skilled knack of doing.

"If it were not so, do you not suppose that your Uncle would have told friend Laurencin to give you only a set sum? Of course, he would have. He is no fool. Yet he did leave the sum unsaid. I tell you, friend, Pierre, this is nothing short of providence."

Putting it as he did, what Bellabre was proposing made reasonable sense to Pierre even though he still had an uneasy feeling about it all.

"Come!" urged Bellabre. "We must secure that which you need. Methinks things could change if we do not hurry."

Once ashore again, the two Pierres went straight to Laurencin's shop.

They entered and properly greeted the merchant. Then Bellabre pulled the Abbot's letter from his pocket and handed it to the man.

"Here, friend Laurencin, is a letter from the Abbot of Ainey, who is a good and generous man." said Bellabre with a confident smile.

The man opened the letter and studied it for a long moment. He had engaged in dealings with Pierre's uncle before and knew the Abbot's handwriting well. Satisfied that the letter was genuine, the merchant looked back at the two young men. "How may I be of service?"

"The Abbot" began Bellabre smartly, "wants to help his favorite nephew, who stands with me before you even now. In as much as it would please the Abbot, it also would gain favor with the King, himself, if you would but comply. This young Squire is also one of His Majesties court favorites."

"What is it you require?" asked Laurencin as he studied Pierre closely.

Bellabre was not taking any chances and poured on the accolades rather thickly, even for Bellabre. "In as much as this young man is a favorite with both afore mentioned gentlemen, methinks it would be prudent of you to aid him quickly."

Laurencin was trying not to become irritated although it was evident that he would just as soon have Bellabre get on with the buying and not talk so much. "Your request, young Sir?"

"My dear friend, here, wishes to engage Sir Claude in a contest of arms and his Uncle has most generously consented to allow this to happen." answered Pierre's companion.

"I assure you, gentlemen, that there is nothing I have that is not at your disposal" pledged the merchant as he, no doubt, was already counting the profit he was just about to make on his merchandise.

"Then Sir Merchant." sang out Bellabre happily. "The Abbot does wish to have his favorite nephew well prepared for his venture against Sir Claude. "So he shall tell you what he doeth require."
Urging Pierre to be prompt in his selection, knowing

that the Abbot would soon realize his great blunder, Bellabre continued the press.

"We must hurry, my friend. Time is money you know. We must make haste and be gone so this nice gentleman can get on with his business with others."

Meanwhile, back at the Abby, the old Abbot had been pondering all that had taken place with his young nephew. Suddenly, growing ashen faced, the old man stood and pressed both hands against the sides of his head. "On my faith." cried out the Abbot with a great deal of panic in his voice. "What have I done?"

Turning to the monk, who was seated nearby in the garden with him, the Abbot said urgently, "friend, Nicholas, I have made a most distressing blunder. In wanting to help my nephew in what is sure to be a mishap, I did present his companion with a letter to the merchant Laurencin for to hand over to him all that he should ask for."

"Of a truth, that was a most kindly thing for you to do." replied the Monk.

"No, no." argued the Abbot. "You do not understand. I foolishly forgot to write in a limit to his spending. That lad could well spend two thousand crowns and that good merchant would not question it. Hasten, this very minute, to reclaim the letter and remind Sir Laurencin that no more than one hundred and fifty francs is to be given to my nephew."

The Monk, of course, set out promptly to do the Abbot's bidding. However, he was much too late in his endeavor. When he arrived at the merchant's shop and inquired about Pierre, he found that the young Squires had been there but had already gone.

Knowing that he did not really relish being the bearer of bad news to the Abbot, Monk Nicholas sighed and asked "Friend merchant, the Abbot wishes to know how much the lad did spend."

"Oh" answered the merchant with no small amount of pride in his voice. "You can safely inform the Abbot that I did take very good care of his beloved nephew, just as he requested me to do in his letter."

"How much?" screamed the very nervous monk. He actually didn't want to hear the answer, which he was sure was going to be most displeasing to the Abbot.

"Just under a thousand francs" answered the merchant.

As the merchant's words burned deeply into the Monk's memory, the man of the cloth shook his head, crossed himself and rolled his eyes toward the heavens. "Oh, by Our Lady, you have spoiled it."

"But" stammered a confused Laurencin "I only

followed directions"

"I cannot explain now, friend Laurencin." answered Father Nicholas who was, by now, in a state of near panic as he turned and headed toward the door. "It is not your fault". Then, talking to himself, the monk moaned aloud "Oh, dear, how am I going to tell my lord, Abbot about this?"

After the monk had left Laurencin's shop, the merchant shook his head as he eyed the door. Then he muttered aloud "but the lad appeared to be most capable." Laurencin closed the door and shook his head again. "Sometimes these church folk can act truly strange, methinks."

Nicholas hurried back to report to the Abbot, who was taking his evening meal.

"Ah good Nicholas." called out the Abbot upon seeing the monk enter. "You have returned. I trust the matter has been settled."

A very nervous Monk Nicholas tried to avoid looking straight into the Abbot's eyes as he rather sheepishly answered. "I did see the merchant as you instructed."

Noticing the hesitation of the monk, the Abbot wiped his chin of some stray broth and cocked his head to one side; a habit he had cultivated whenever he was becoming annoyed. "And? Pray tell me. What is your delay?"

"I was too late for he had already supplied your nephew with all that he asked for," reported the monk nervously.

The Abbot rolled his eyes to the heavens and with much hesitation asked the question he really did not want to hear the answer too.

"And the sum of it all?"

"It was just under a thousand francs, Lord Abbot."

The Abbot nearly choked on his food at hearing the monk's answer.

"A thousand francs?" he screamed. "A thousand… Holy Mary. Go find where my nephew does lodge and inform him for me that if he does not give it all back…."

The Abbot searched for the words he wanted and finally finished his sentence. "If he does not then he shall not, in the future as long as he lives, ever receive as much as a farthing from me."

CHAPTER SEVEN

Poor Monk Nicholas attempted to do as he had been instructed to do. However, the fast thinking Bellabre had foreseen just such a move and, upon returning to Lyons, had instructed all he could talk to, to stall anyone coming to ask questions.

After more time misspent in returning to Pierre's quarters, Nicholas was promptly informed that the busy Squire had accompanied Captain d'Ars on a mission for Count d'Ligny and would not return for three days to meet Sir Claude. That was to take place the following day and after that, it would be far too late for anything to be done concerning the Abbot's great blunder.

The steward did arrive at the place he was told that Pierre lodged but was informed that the Squire had gone to the House of d'Ligny. Nicholas made his way to where he now thought Pierre would be only to be informed that the Squire did not live there.

It was a very weary and dejected steward who had to report back to the Abbot about his failed mission.

"Well, friend Nicholas?" inquired the Abbot anxiously. "I trust that my nephew saw the good sense of it to return what he took."

Nicholas gulped hard and, with a sheepish tone to his barely audible voice, uttered the words which he well understood would bring on another outburst from the Abbot.

"I tried all day long to catch up to your nephew." reported the defeated monk. "But, alas, he was as a shadow and could not be found."

For a very long moment, the Abbot simply sat trying to digest the bitter words that the steward had just uttered. Then, very uncharacteristic of him, instead of flying off the handle as was expected, the Abbot took a deep breath, stood up and slapped his thighs.

Perhaps, in the back of his mind he was mulling over many facts, including the fact that if it hadn't been for Pierre's family support, he would not be in the position he held in the first place. Perhaps he decided that to make more of the matter would, in some way, sully his name with the community. At any rate, even though he did not like what the two squires had done, the Abbot decided that to pursue the matter would cause him far more problems than the coins he had already lost. The Abbot simply, to no one in particular, remarked aloud. "He has been a bad boy, to be sure. However, being of good stock and breeding, I am most confident that he will repent of his fowl deed and regain the good graces of Our Savior. No more shall be heard of the matter."

"As you wish." quickly agreed Nicholas.

Neither Squire got much sleep that night. All preparations had to be quickly put into place for Pierre to be able to live out his dream. Bellabre officially carried through the motions for Pierre to enter the contest while the lad from Bayard busied himself with the last minute preparations of his new found equipment and battle chargers.

As was the custom on such an occasion, Pierre spent a respectable period of time saying his morning prayers.

The day was abuzz with the excitement of nine gallants and their challenges to Sir Claude de Vaudrey. Some even quietly scoffed at seeing Pierre's name among the challengers. Those, of course, were mostly people who did not know the young Squire personally. All who did know him knew, deep somewhere in their souls, that with this brave and accomplished young stalwart anything was possible. Some of the more experienced soldiery, though wishing the young Squire well, had to commit to their experience and knowledge of the great French champion and think of the venture as folly.

"I hope I have not bitten off more than I can swallow" admitted Pierre privately to Bellabre.

"What is this?" snapped the other Pierre. "Is Bayard having doubts? The very thought of it makes me laugh. You are very accomplished and I wager are every bit as good as Sir Claude was at your age."

"Yes, but that is where he started out" confirmed Pierre with a slight reservation in his voice.

"And he is still very good" admitted Bellabre with a huge smile. "But you have youth on your side and my wager is on you, friend Pierre."

Pierre had to admit that his new best friend was a master confidence builder. Bellabre's words stuck with Pierre for the next hour as they joined the other lances that had made the bold strike upon the champion's shield.

The crowds began to gather early as this was a rare event and enormously popular.

The nine challengers were led to the lists by their squires, where they quietly awaited for the festivities to begin. King Charles and his huge entourage, according to proper ceremony, filed in and were seated. Then the celebrated man of the hour, Sir Claude de Vaudrey entered to his accorded applause.

As Pierre sat there, face covered, at the end of the line of challengers, he observed the men who would be facing Sir Claude. The first two were Scottish wanderers, neither of whom he had seen before. One was young and seemed very nervous while his companion was an obvious veteran of such games and sat quietly.

Next to them sat two Swiss knights, both of whom Pierre had seen in and among the king's soldiery. Pierre, however, had not had occasion to deal with either thus far.

The next four gentlemen were some of His Majesties finest lances. There was young Benneval followed by the Seneschal Galliot, who was a leading master of Artillery. Next to him came Chatillon, a rich and spoiled but sufficiently accomplished knight.

Next to Pierre sat a young, moderately accomplished knight named Sandicourt. Actually, it was no surprise to see him there because he was one of the most passionate tournament fighters in all of France. Just recently he had all but gone completely broke giving a splendid tournament that would be known for years as Pas d'arms de Sandicourt.

Pierre, the young squire from Castle Bayard rounded out the group of challengers.

Because, according to the custom of the event, all the challengers' faces were covered by their visors,

Pierre couldn't actually see their eyes. Perhaps that was a blessing in disguise, he was thinking, as they could not see his eyes either. Only a handful knew who Pierre was as he had no armorial markings, being not yet a gallant to have golden spurs.

Pierre actually preferred it this way as he felt it might give him somewhat of an advantage against this most seasoned warrior.

Prompted by King Charles, Sir Claude made his short but most elegant speech of how he had received permission from His Majesty to hold this event.

Amid a myriad of cheers and much hoopla, the challengers were recognized, en masse, to take their proper bows.

Then the fun began.

Although both of the Scottish knights seemed well enough skilled, neither proved much of a match for the seasoned French champion. As was the custom, after each of the challengers had completed their turn, they were led, head bare, before the lists so all could see who had done well or not.

The two Swiss officers met just about the same fate as the men before them as it was evident that Sir Claude was at the top of his game and having a great time.

Then both Bonneval and the Seneschal kept their horses but were so bruised by the champion's blows that they declined the meeting on foot as was their right.

All eyes fell upon Sandicourt next. He was well known and extremely popular. More than a few onlookers were pulling for him to, somehow get the best of Sir Claude.

Some in the crowd were even looking beyond Sandicourt as they speculated about the final challenger. He had no armorial markings and few knew anything about him. Some reasoned that he was a foolish lad who, even though had displayed skills beyond his years, had bitten off more than he could handle. A few, a very few, even voiced the opinion that he was naught but a brash and cocky man who was simply seeking glory.

However, all who really knew Pierre well understood that nothing could be further from the truth. Pierre genuinely was seeking only the experience and knowledge to be gained from the encounter. Nobody could adequately describe to the young Squire what it would feel like to go against the champion of all of France. Only the doing of it, itself, would satisfy the man from Bayard. Some were even overdoing it, saying Pierre could unseat Sir Claude. Pierre, on the other hand, was considering it would be a grand coup on his part if he simply did not lose his saddle of the first pass as most of the others had done.

But, first Sandicourt had to take his turn.

The loud cheering cut deep into Pierre's thoughts and jarred him back to the reality of the moment.

Young Sandicourt had actually handled himself well against the French champion. Emulating Bayard, himself, the young lance had turned in his saddle just enough to not absorb the full brunt of Sir Claude's lance thrust. However the young knight reeled in his saddle and toppled, haplessly, backward. His landing was rather comical; enough so to prompt loud laughter from some of the onlookers.

Normally, that would have been the end of Sandicourt. But Sir Claude, who seemed to be enjoying this particular event, for whatever reason, decided to give Sandicourt a chance to redeem himself by some small measure and invited him to meet on foot with swords.

Sandicourt well understood the great honor being extended to him. Not wanting to appear ungrateful, he quickly drew his sword and gamely faced de Vaudrey. Although his skill was passable, the young lance soon realized that he was far overmatched in this case. After only nine sword blows from Sir Claude, Sandicourt knew that the battle was lost.

His smile was huge, though, as he was led along the lists to take his well deserved bows.

Pierre looked at Bellabre and took a deep breath.

This was it, he was thinking as he heard the herald call out to him. This was the time he had been waiting for; the chance to ride against the best in the land at the moment. Somewhere in the deep recesses of his mind, Pierre was also considering that he just might have overstepped himself in this case. On the other hand, though, the young man from Castle Bayard had developed an uncanny awareness concerning his own very extensive prowess in all areas of weaponry.

He genuinely felt that he had as good a chance as any in the lists to beat Sir Claude.

At any rate, here he was; a young squire facing the most accomplished tournament fighter in the country. Pierre crossed himself and uttered a small silent prayer as he guided his mount into place at the end of the list.

Both he and Sir Claude saluted King Charles and then turned and saluted each other. The signal was given and lances were lowered as the crowd of onlookers buzzed loudly about what was taking place.

The two powerful battle chargers lurched forward.

Pierre concentrated on two things in that brief

moment. First he became keenly aware of Sir Claude's blunted lance point, which would surely be aimed at just the right spot on his shield. The other thing Pierre became fixed upon was his own target.

Through his brief experience at jousting with the lance from horseback, Pierre had formed the opinion that if his lance point struck the other person's shield just at the bottom, the force had a double effect. First, it caused the said shield to bend inward with no small amount of pressure.

Sometimes the lance, in striking low enough, would actually slide down from the shield and into the man's body. The second thing was that, either way, the other person received the maximum shock from the blow.

Sometimes, too, the lance would strike a telling blow just below the shield, which would hit the desired target and almost always unseat the foe.

Both riders grunted at the force of the other's lance strike on their shield. Both lances splintered as the crowd erupted in uncontrollable applause.

Pierre felt the shockwave move all the way up his arm and into his shoulder as he felt himself reeling in his saddle. He, however, was much too good of a horseman to lose his saddle at that kind of a blow. The lad from Bayard slid, awkwardly, to one side of the saddle and then quickly righted himself as if being pulled up by unseen strings. The very appreciative crowd erupted with great applause at this. Even Sir Claude had to be a little impressed, thought Pierre as he turned his mount at the end of the list to, again, face the champion.

Pierre confirmed the respect that he knew he would have for his foe. But, he couldn't help but believe that his pass had gained the full attention of Sir Claude, as well, if not a good deal of respect.

Of course, Pierre was only a passing name to Sir Claude, who had no way of knowing that the rider he now faced was only a squire, and a new one at that. He only understood that he had a real challenge before him. Sir Claude signaled that a second pass was in order and the crowd buzzed with various stages of wonderment and anticipation.

Who was this unknown person that was showing himself so admirably against the great champion?

Again, the two riders streaked toward one another. Again, both lances shattered upon impact with the two shields.

This time, amid the complete madness of the very appreciative crowd, Sir Claude turned his mount, gazed back at his foe and then offered a rare salute of complete respect. Pierre returned the gesture and then saw the champion motion for him to dismount.

In the stands, King Charles became a bit concerned as he turned to Count d'Ligny and spoke.

"Cousin, Messire Claude surely realizes that it is a lad he is facing."

"I do not think he does, Sire!" answered the Count.

"Then pray, should we not stop the fo... foot assault be... before the lad gets hurt?"

"Of a truth, Your Majesty," countered Nobleman, "the lad is of unparalleled skill. Methinks it would be a most interesting confrontation. Sire Claude will not hurt him as it is a friendly contest. Pray let it continue."

"As you wish, cousin." conceded the king. "You understand more of these matters than I."

On the field, Pierre and Sir Claude faced one another as the champion called out.

"My compliments, sir knight. Your riding is excellent." said the man. "Now I want to see your swordsmanship."

Pierre saluted the champion but said nothing although he was thinking "and I, sir, have long awaited this moment."

For the first minute of the cat and mouse game, each contestant circled and simply tested the other's movements. Sir Claude was the expert of the day and Pierre studied his movements carefully; wanting to learn anything that might help him in his own cause. The champion quickly realized that he was not up against the run-of-the-mill soldiery in facing this foe. The knight with no markings moved like a cat on the prowl. He had already parried six or seven of the champion's blows with abnormal ease. This, alone, began to intrigue Sir Claude. He had been at this game for many years and seldom had confronted anyone who even stood against him, let alone gave him any type of competition.

There followed an exchange of six or seven more stalwart blows from each man. In using all of his cunning and strength, Pierre had succeeded in matching Sir Claude at every turn.

On Pierre's final block, his blade slid off Sir Claude's sword and struck a nearby post so hard that it broke in half.

"Hold!" cried the Herald and three of the umpires almost simultaneously.

They had witnessed a marvelous showing from both contestants and decided to stop the affair before anyone got seriously hurt.

The onlookers were beside themselves with applause and noise to show appreciation for what they had just witnessed.

In the Royal box, King Charles fairly bubbled with excitement.

"On my s… soul, Cousin" he called out. "I do think tha… that Picquet has had a ma marvelous beginning of it."

"You speak well, Sire!" agreed Count d'Ligny with no small measure of pride in the young man from Bayard.

"Methinks" continued the king "That I nev..never gave so good a present to an…anyone as when I gave him to you."

"Sire!" said the Count. "If he grows to be the great man that I think he will, then it will be more to your honor than mine. For it is due to all of your kind praise that he undertook this venture to begin with."

On the field, Pierre and Sir Claude both removed their helmets. The French champion was slightly stunned when he discovered that his very formidable adversary was a mere lad who hadn't had his first shave yet.

"By my faith." exclaimed Sir Claude. "Your deed was truly amazing. You remind me of a young … me. By what name are you called?"

"I am Pierre Bayard." was the answer.

Sir Claude then, very uncharacteristic as his great ego generally stood in his way on such occasions, reached down and took Pierre by the arm. The Champion raised his foe's arm high and called out to the gallery.

"Pierre Bayard. None, in all my days have I ever met with more skill."

The crowd erupted with delirious applause.

A few moments later, the diminutive Monarch declared a supper to be held in Pierre's honor and women and men, alike, lined up to merely shake his hand or give him a hug.

CHAPTER EIGHT

"Squire, assist me with my horse."

The authoritative voice seemed vaguely familiar, yet strange, as it barked out the command. Upon realizing that only a knight or some other important personage would be using such a command voice, Pierre turned quickly; ready to obey.

The Crimson Shield with the Blue Dagger seemed to engulf the entire field of his vision for a brief moment. Pierre was taken by slight surprise at seeing the crest which he had developed quite a disliking for.

Pulling himself together, Pierre watched the knight as he quickly dismounted and turned to hand him the reins of his horse.

"I am Sir James," announced the knight, "the new lieutenant of the Sixth Company of the lances of the House of d'Ligny."

The knight paused as his brows narrowed while taking a closer look at the Squire standing before him. Then recognition set in. "You? See that he is curried and fed." The man looked hard at Pierre then added. "Squire!"

The brash young knight said nothing more. He simply turned on his heels and left. Sir James, thought Pierre as he watched the knight walking away; the man who had acted so boldly toward Anne at the tournament and made disdainful remarks about her Page. Well, thought Pierre, I am neither young nor a page any longer.

Once more, Pierre's training schedule was increased as a squire was expected to do all things faster and infinitely better than ever before.

Pierre picked up as a Squire where he had left off as a Page by excelling in every challenge that was put before him. Everyone, even Captain d'Ars, was saying that he was the most natural candidate for knighthood that he had ever seen. Pierre served, several times, as 'Squire of the Body" to Captain d'Ars and the count, himself. He also served several other men-at-arms and accompanied them on various missions for the count. This business included tax collecting, chasing thieves and putting down small rebellions.

Soon Pierre Bayard was being recognized and promoted as possibly the finest squire in all of France. Bellabre was getting his due, as well, but it was Pierre that was the apple of everyone's eye. Pierre had grown and was now standing at six feet and weighed one hundred and ninety five pounds.

His skill with all his weapons matched or surpassed most of the knights, themselves.

Many remarked as to how they dreaded the day that Pierre would gain his golden spurs as he would, then, rule the tournaments and take all the prizes.

Even now, only Pierre Bellabre and three or four knights were in the same league with him in most cases.

So many people of important stature constantly visited the extraordinarily powerful and popular Count d'Ligny that Pierre began to seldom give notice. One day, though, a visitor did come who commanded the full attention of the young man from Bayard.

Count d'Ligny and Captain d'Ars were giving a gentleman with a very familiar face to Pierre a tour of the grounds. They came upon the two Pierre's who were just finishing up a light weapons practice. After watching a few moments, the visitor called out "My compliments, young sirs. I take particular delight in watching good swordsmanship as it is my greatest field of interest."

"I trust you gentlemen know Messire Claude de Vaudrey" spoke out Captain d'Ars.

"We thank you most kindly for your words, my lord." said Pierre as he tried not to stare at the great champion of France.

"Messire Claude." said the count. "May I introduce you to two of the finest squires in the land? This silent one is Pierre de Poquieres, de Bellabre et de la Marche."

"I salute you, squire." said Sir Claude. "Your combat shows that you have been paying diligent attention to your training. Believe me, young sir, that is the secret to it all."

"Thank you, my lord." countered Bellabre with a smile of satisfaction.

"And this, my lord," continued d'Ligny with no small amount of pride, "is Pierre Terrail of Bayard, who is perhaps the finest horseman in all Europe, be he man or boy."

Pierre felt honored as Sir Claude looked into his eyes and put his hand on Pierre's shoulder and spoke. "Of course, we have met and I know his family well. And, of a truth, I have been delighted by this young man's riding before." Then speaking straight to Pierre, the master swordsman continued. "Your expertise with your weapons nearly rivals that with your steed, young Bayard. Methinks I might retire. For before you are done of it, Sir Bayard, your name will be much more famous than even my own."

"Your lordship is far too kind." said Pierre. "You put me in far too gifted of company. However, I truly thank you for your thought of it."

"Fare ye, well, young squire." said Claude. "Fare ye, well."

Claude de Vaudrey nodded toward Bellabre then gave a small wave as he and his hosts turned and continued on with their walk.

"He likes you, my friend." reminded Bellabre. "He really likes you."

"He liked both of us." said Pierre.

"And why not." retorted Bellabre, who was feeling good at the man's words.

"Are we not two of the finest squires in the land? After all, it was you who did so well against this much accomplished gallant."

"Do not let your head grow too large, my brother" reminded Pierre playfully. "This is a big land."

"You, squire from Bayard. Hold!"

The voice had the rude tone that Pierre had not heard many times but had learned to recognize instantly.

Both Pierres turned to see Sir James striding toward them.

"It may be a big land, mon ami" said Bellabre in a whisper. "But this is becoming a small courtyard."

Sir James, as might be expected, had obviously become jealous over the accomplishments of the young Squire as had been noted almost daily around the household of d'Ligny. The knight seemed to delight in going out of his way to try to make Pierre look foolish before others. The knight seemed to resent the fact that everyone, including the ladies and some of his own peers, spoke more highly of a squire than they did of a lieutenant of the lances at the court of d'Ligny.

Sir James threw an arm load of armor and bootery on the ground in front of Pierre Bayard.

"These need a good polishing, squire." barked the knight. "See to them and then deliver them to my quarters before you dine tonight."

Captain d'Ars, who had been following a few yards behind the count and his guest, had overheard the young knight's conversation with Pierre. With a solemn face and understanding the situation between the two, d'Ars walked back to where the two were facing each other.

"Well, don't stand there like a dote." said Sir James impatiently. "Pick them up."

Pierre bent to begin retrieving the items when he heard another hearty voice call out.

"Hold!"

The two Pierres and the knight all looked to the speaker as Captain d'Ars walked upon the scene. The captain turned to look at the knight. "If it be that you are in need of a squire, I'll gladly send you one, lieutenant. But, alas, these two belong to me and I have more than enough duties to keep them busy the day long."

"Both of them, my captain?" questioned Sir James.

"As I said" answered d'Ars, "I have much to keep them busy with."

Sir James did not relish the idea of the squires being let off his hook but he thought even less of the idea of crossing the likes of Captain d'Ars.

"My captain." said the knight before turning to walk away.

"Sir knight?" called out d'Ars, toying with the man now. "Your belongings?"

"Leave them, sir captain." called back the knight. "I'll find a squire to attend them."

Captain d'Ars and the squires watched the knight stomp away. "There goes a young man who could use a little attitude adjustment." said d'Ars quietly. Then he looked over at Pierre with a wry smile. "Methinks he should be content that you are not, yet, possessor of your spurs, Sir Squire."

Pierre, of course, said nothing aloud but inwardly agreed with what the captain said.

Captain d'Ars nodded his 'goodbye' and turned and left.

"Of a truth." said Bellabre. "What he said was right. It would be a lesson that he would not soon forget. Someday it will come to pass."

"Perhaps, my friend" agreed Pierre. "But right now we have work to attend too."

For a few months more the hard work and learning continued for Pierre and the others. Count d'Ligny found himself particularly troubled by an upstart and annoying bandit band which had been roaming freely in the outer hamlets of his domain and harassing and bothering the people.

Deciding to put an end to them, the count ordered three of his knights to take the matter into hand. It was only a matter of time until circumstances would have pitted Pierre and the knight with the crimson daggered shield against one another

This was the time. Pierre was assigned as Sir Jame's squire. With Bellabre off and serving another knight in another venture, Pierre found himself more or less alone as he did not know the other squires all that well.

"Just so there are no misunderstandings," began Sir James in his usual unpleasant way at the meeting before they started on their mission. "I am in charge and you are my squire. I expect you to obey my every command to the letter."

"Understood, sir knight" answered Pierre in as pleasant of a voice as he could muster. After all, he reasoned, Sir James was a man-at-arms and Pierre was a squire and they were in the same service and on the same side. It would be foolish to continue to feud. Besides, said Pierre to himself, I have more important things to do than to carry on a feud with a knight with no apparent reason. What had the man done except to think that Anne was beautiful? She was, so he could not really blame Sir James for that. So, Pierre decided to lay aside his ill feelings toward the knight.

Thus, they set out; three knights, four soldiers and four squires all going bandit hunting.

For three full days the small group tracked the bandits. Pierre made life on the trail a bit easier with his skill at setting snares and they ate well.

It was a quiet night. This was the fourth night on the hunt. Pierre had just been relieved of sentry duty and was spreading his bedding beside the cozy fire when he froze in mid motion at the sound. The frightened whinny of a horse and a warning cry from the squire on sentry duty suddenly brought the camp to life.

The count's soldiery had been taken a bit by surprise as confusion and bedlam seemed to be the order of the moment.

Grabbing their swords, with no time to get into armor, the knights and soldiers leaped to their feet to face the attackers. The squires were instructed to take cover as the large bandit force attacked.

Confused and frightened at their first battles, the squires ran for the woods. All except Pierre, that is. He had not trained all these years to take flight at the first sign of danger. Acting on instinct, Pierre grabbed a nearby shield and smashed it, hard, against the shins of the nearest figure in the dark. The bandit yelped in pain.

Then Pierre's strong hands ripped the man's sword away from his grip and then sent it almost to the hilt in the man's gut. That next scream was a death scream and it was a sound that slightly unnerved the squire from Bayard. Then Pierre lashed out at another nearby figure and there was a second scream of pain and death.

In the next instant, two or three pairs of strong hands had Pierre. He was knocked to the ground. He felt a vise like grip on his own throat as someone had his arm locked tightly around his neck. Another man knocked the weapon out of his hand with a club. Then the fighting squire felt the ground come up and hit him in the face about the same time that he felt the pain from the crack on the back of his head. He did not feel his body rolling down the hillside in the darkness. As a wave of blackness swept over him, Pierre thought of the Quintain.

Pierre could feel the dirt on his tongue and the

searing heat from the sun beating down on his face. He spit and wiped his mouth on the back of his sleeve. It was a movement that only caused the throbbing in his head to feel worse.

Pierre tried, awkwardly, to sit up and found himself in some thick bushes. He suffered a new wave of pain as he dragged himself from the brush and crawled up the small embankment.

Topping the rise, Pierre immediately let his mind reconstruct the happenings of the night before. The camp was a complete shambles as the bandits had taken everything that they thought would be of any value.

Pierre saw the three prone figures in familiar clothing They were the bodies of three of the soldiers. Pierre realized that the only reason he probably was not dead, too, was because he had fallen out of sight or reach down the embankment. Five other bodies were strewn over the area. They were bandits, two of which Pierre had dispatched. Using his foot, Pierre turned one of them over and was pleased to see the man's sword had been trapped under his body. The Squire quickly picked it up and stood, letting his gaze make a quick sweep of the area.

The Crimson Shield with the blue dagger was buried halfway under some brush. Retrieving it,

Pierre stood studying the ground. It looked like about two dozen bandits had jumped them. They had taken all the others as hostages to use as whatever bargaining chips for whatever they could get as a ransom.

The trail led off to the east and Pierre had no choice but to follow. Perhaps, there would be something that he could do to help the captives. His one big advantage was that he had surprise on his side and none would realize he was even alive, let alone think that he would try to follow them.

Fighting the dull aches in his body, Pierre had not traveled far when a snap of a twig brought him spinning around and ready to meet whatever danger there was confronting him. Pierre sighed in relief when he saw the other squires. They rushed toward him, all talking at once. Pierre called for silence. "Stay calm and tell me what happened" he said.

They related to him that the bandits went to the east and that they did not think they were too far away.

"Good!" said Pierre. "Maybe we can catch them and see about helping our gentlemen."

"But we have no weapons" protested one squire.

"Fete Dieu" scowled Pierre. "I care not a fig of it. We have our wits and we have surprise at our backs. They will be feeling too cock sure of themselves to be on the alert. They would never consider that a group of squires would hunt them."

Seeing the squires' looks of bewilderment Pierre barked quietly. "Look, is this not exactly what we have been training for? We are Squires, not helpless Pages so let us start acting as such. We have a huge advantage of it so let us take advantage of what we know."

The Squires shared knowing looks of sheepish self-scolding as they knew that what Pierre had reminded them of was the truth.

The Squires took off, on foot, following Pierre, who ran awhile then walked and listened awhile. As the sun sank from sight in the western sky, Pierre called for a halt.

"The tracks say they are near" pointed out Pierre. "They will be making their night camp and I feel we will come upon them soon."

The same squire, the youngest, that complained before, did it again.

"But we are only squires and armed with but daggers."

"What you say may be true" countered Pierre, "but we are French squires and from the house of d'Ligny. What are they but rogues and bandits and undisciplined?"

If Pierre's answer did not make much sense to the squires, it at least inspired them, because they took heart and actually began to think of ways to take on the bandits once they caught up with them.

After a twenty minute rest, the squires struck out. Soon, through the darkness, the tiniest bit of light could be seen through the forest.

"There they are" said Pierre quietly. "Only a few hundred yards ahead. On my faith, we have them. We must move as silent as shadows and when we strike we must do it with the fury of ten of our number. Does everyone understand?"

The other three nodded. "How do we attack?" asked one.

"Remain here while I take a closer look" said Pierre. "We will know more after I see how they are deployed."

Soon the squire from Bayard had crept close enough to get a good look at the bandits' camp. A careful count set the number of bandits at a dozen. The captives were bound, back to back and laying on the ground away from the roaring fire. The horses were all bunched together and had been tied, loosely, to shrubbery and trees.

How careless of them, thought Pierre as he eased himself away from his hiding place and hurried back to the waiting squires.

"We are in luck" reported Pierre. "They are but twelve and have posted no sentry that I could see. They will be going to sleep soon."

"That is what we need" said one of the group.

"Exactly!" agreed Pierre. "Try to rest. I will awaken you in about an hour."

CHAPTER NINE

Two hours later, enough for the entire bandit camp to be fast asleep thought Pierre, he awakened the squires. He went over a simple but effective plan of assault.

"Each of you find the strongest piece of wood you can and use it as a club when the time comes. Wait for the call of the owl before you do anything" warned Pierre. "If you do not wait for me you could give us away and we would be captured or maybe killed."

After double checking with each of the others as to their prescribed duties once they reached the destination, Pierre led the squires through the night.

The fire had nearly gone out. This made it easier as there would be no bright light to give them away, thought Pierre. Pierre slipped up behind one man who was sleeping against a tree. Pierre's dagger, properly placed in the strike, rendered the man unconscious and Pierre eased his body to the ground.

With hand signals, Pierre directed the squires'

movements. In his anxiety, one of the squires almost stepped on a sleeping bandit, who had suddenly rolled over in his sleep. Everyone froze in mid step until Pierre gave the signal to continue.

Pierre's second well placed pebble on Sir James' ear awakened the knight without disturbing any of the sleeping bandits. Sir James quickly alerted the other equally surprised knights that help was at hand.

Pierre moved quietly to the prisoners and quickly cut them free. Then he nodded to the other squires, who were waiting near the fire for his signal.

At the signal, the other squires moved into action. One squire went for the horses while the other two, immediately dumped a blanket full of dirt on the embers. The camp was pitched into total darkness.

After the squires placed some hasty but well aimed blows on as many nearby targets as they could reach, the rescue party and the knights made a dash for the horses.

After mounting Wingfoot, Pierre kicked out at a figure nearing him in the dark. By the time the sleepy bandits gathered their wits, the squires and their rescued knights were long gone with the remaining bandit horses scattered in all directions in the dark. The bandits stood around looking at each other in the dark without a clue as to what had just taken place.

Sir James called the group to a halt about a mile from the bandit camp. The knights seemed stunned at the bravery and daring of the squires. They had not expected to be rescued, at all, let alone by a group of squires. The other squires were very quick to point out that the entire plan was Pierre's idea and that he acted no less bravely than any knight would have acted in his place.

The knights thanked the squires and commended them, again, for their bravery.

Even though it was apparent that Sir James had great difficulty in doing it, the knight with the Crimson Shield directed his words to Pierre. "I owe you, squire. It was a brave and well executed thing you did."

Although the young knight recognized Pierre's accomplishment, Pierre could still sense some hostility from the man. Pierre knew that Sir James was grateful for what had happened but, at the same time, the knight resented it as the deed would only add to the growing glory of the Squire and serve only to enhance his already considerable position in court.

When they returned to the court at Lyons, Pierre's deed was told by all the other squires. Everyone was justifiably proud of the young man from Bayard. Count d'Ligny, immediately dispatched Sir James, who wanted revenge, and his entire company to handle the bandits. In an added surprise, the count also issued an edict announcing that there would be a grand feast and ball held in honor of the court's favorite squire.

"My brother." said Bellabre proudly. "Upon my soul, you are, indeed, the most honored and best known squire in all history."

All of this, though, served only to build Sir James' resentment against Pierre. He found various opportunities to try to embarrass or belittle the squire. Even though Pierre did not enjoy being made sport of, he cherished his honor above all else. So, he kept his temper no matter how badly the young knight treated him.

However, the knight began to tread on thin ground when he made a couple of unwarranted remarks concerning Anne and her 'boy'. Sir James was fortunate that Pierre had only heard of the remarks rather than having heard them directly. If the latter were true, knight or no knight, the squire of the realm would have been obliged to take Sir James to task.

More than once, Pierre had absently remarked that the man needed a lesson in manners.

"To be sure, you would be the one to do it, mon ami" said Bellabre. "But even you cannot just confront a knight with a challenge. Besides," laughed Bellabre, "this is no time for fighting. This is a time for dancing and there are wenches to be had."

"Do you always think of nothing but women?" chided Pierre.

"Cre'Dieu?" smiled Bellabre. "Is there anything else?"

The party and dance lasted for hours as everyone was having a grand time.

Pierre spotted Count d'Ligny and Captain d'Ars and a couple of others in a serious huddle in a corner and wondered what the secrecy was about. He did not have long to wait for an answer.

Paul of Luxembourg finally drew everyones' attention as he wrapped on a goblet with a knife.

"Tis truly a festive occasion" he called out when the noise subsided. "For it does great honor to a brave young heart of France. It is rare that a man such as this appears among us and we are fortunate that he is in our court. In this case, that brave young heart beats in the chest of the finest young squire in this or any other land."

The crowd cheered as Count d'Ligny held his huge goblet aloft. "I give you a toast, Pierre Bayard." he added in a loud voice.

Immediately, every cup was raised in honor of

Pierre followed by a thunderous roar of applause. The count was joyously shouting "Picquez, Picquez!"

Then all those in the dining hall echoed the name loudly.

Count d'Ligny held up his hands again, demanding silence. Then he motioned for Pierre to come before him and kneel. Pierre did so, not understanding why but willing to comply with whatever his master wished.

"You have brought great honor, not only on yourself, but on the House of d'Ligny and all of France, as well."

As Pierre knelt there with eyes closed, a million things went through his mind. He felt a bit awkward at all the attention, but then he always felt awkward in the lime light. He scarcely breathed as he felt Count d'Ligny's garment brush his cheek. He almost jumped out of instinct when he heard the sound of the steel blade being drawn from its scabbard.

Suddenly, Pierre Bayard realized what was taking place and he felt the same kind of warmth come over him as he did when Anne first kissed him.

Paul of Luxembourg again was speaking. As the Count was speaking, Pierre could not quite believe what he was hearing.

"You, Pierre Bayard, have exemplified yourself

in all that you have undertaken, whether a large or small task. Your skills match and, indeed, sometimes surpass those of many men-at-arms. Your recent bravery in the face of the enemy and the subsequent results thereof are more than admirable."

Pierre felt a great lump of pride in his throat as he felt Count d'Ligny's great blade touch lightly on his shoulder as the count continued. "You have grown to be the epitome of all that embodies knighthood. I, therefore, strike thee once, twice, thrice, dubbing you Knight of the Realm, Pierre Terrail Bayard. Arise, ye, Sir Knight."

While every voice in the great hall screamed with delightful approval, Pierre tried gamely to hold back his tears as he forced himself to stand. He wanted to say something, yet, had no voice. Then he felt the tears of overwhelming joy flow down his suntanned cheeks and he thought of Anne and of his family and wished they were there to share this great moment with him.

Pierre looked at Count d'Ligny, who was bending over doing something before him and then over at the proud faces of Captain d'Ars and Bellabre.

Count d'Ligny straightened himself up and, once again, stood before the new knight. This time he had something in his hand as he spoke.

"No true knight of France shall be without his

golden spurs. It would be my extreme honor if you would accept these from my own boots" said the count as he shoved the spurs toward Pierre.

"Oh, but my lord," said Pierre as he started to protest, "they are your own spurs."

"They are small enough reward for so brave a young heart" replied the count. "They are yours from this day forward. Wear them with pride."

"I do humbly thank you, my lord." said Pierre as he felt his fingers lock around the precious gift. "My loyalty will always be with my God, my country and my king."

"Well spoken, Sir Knight." said d'Ligny. Then he raised his cup again.

"Now, Sir Knight. I will grant thee a boon as it is your right by custom. Only speak of it and if it is within my power to give, it is yours to have."

Pierre's gaze found that of Bellabre and both understood what he would ask for.

Then Pierre's gaze found another face in the crowd. The knight with the Crimson Shield and blue dagger was off in the corner. He had been sulking to himself all evening as his mind would not let go of the fact that it was through Pierre's rescue of him that prompted the expedition of Count d'Ligny's timing on making Pierre a knight. It seemed Sir James was simply destined to lose when it came to coming against the man from Bayard.

At first, Pierre thought of ignoring this young knight and simply writing the man's displeasure off to a quirk of fate that he seemed to carry a perpetual chip on his shoulder. But, then, he took a second thought. If someone did not put this cocky knight in his place he would continue to taunt and demean the underlings of the world.

No! Pierre decided that the time was now. It was time that Sir James received a little 'payback' and perhaps learn his harsh lesson along with it.

"My lord." called out Pierre and getting the count's full attention. "There is one thing that I would like at this time."

Pierre caught Bellabre's gaze and each smiled a small smile.

"Name it, Sir Knight." compelled the Count.

"I wish to issue a challenge of single combat jousting to settle a personal matter between myself and another gentleman." said Pierre. "It concerns an indirect attack upon my honor, which I do hold in esteem above all else."

"You need not say more, Sir Bayard." assured the count. He was an adventurous man and was not unaware of the treatment that his favorite squire had been enduring from the moody knight with the Crimson Shield. As a lover of fun and any good contest of skill and anticipating Pierre's wish, the count nodded for Pierre to speak.

Pierre, admittedly, played the moment to revel in his delight as he slowly walked over to stand before Sir James. "This is the man I want to challenge."

"But, my lord?" protested the already distressed knight. "This is but a squire... well, a knight of brand new issuance."

Everyone in the room seemed to be caught up in the moment as they eagerly began to conjure up images of what was to come. None particularly liked Sir James and besides, they thought it would be a good show of these two strong young men going against each other.

Captain d'Ars, a very astute man himself, did not miss this chance to seal what would be not only a personal triumph for his favorite Squire but also for a good lesson taught to Sir James, as well.

"Then, methinks, Sir James," said Captain d'Ars "that facing a new knight should not cause alarm on your part."

The count, too, found the situation quite sporting and seemed well amused at wondering, along with the others, just how well Pierre Bayard could actually fare against a seasoned knight. In any case, the count thought it would be good seasoning for his newest knight, whatever the outcome of the matter.

"The matter is settled" said d'Ligny. "The parties in question will face each other in the lists at the dawn, forthcoming."

"A prize" called out someone.

"Yes, there must be a prize" agreed Captain d'Ars.

Pierre, completely enjoying the moment for the first time as equals in status with Sir James, looked at the seething young man and smiled slightly.

"If it pleases my lords," spoke out Pierre, "prize enough would be that the loser of this contest presents himself in Grenoble before the lady Anne de la Chavet, whose honor has been impaired along with mine own. There the loser must proclaim to the lady in question that he is the lesser champion."

"Bon! Well spoken." said the count as he nodded his head in approval. "Love has no equal, Sir Bayard. With the pass of the lance it shall be done and the first to lose his saddle will be the loser. Do these terms meet with the approval of both contestants?"

By now, Sir James was so angry that he would have agreed to about anything to get a chance at this upstart who dared challenge him. He nodded and then stalked off.

"I claim the right to be your squire in this affair." said Bellabre.

"I would have no other, friend Bellabre." said Pierre with a huge smile as he put his arm around his friend's shoulder.

"He sure was angry" noted Bellabre.

"Good! The angrier he stays the less of a clear head he will have" said Pierre.

"The pompous peacock best pray that you do not remove that head." laughed Bellabre

Try as he would, Pierre got very little sleep that night. He had made the challenge and now he had to carry through with it. He was in chapel very early and then could eat no breakfast. Bellabre, of course, ate as if he were starving. He, mused Pierre to himself, was not the one who would be facing Sir James.

The sun had not finished its early morning climb and the galleries were already nearly packed. This was a spectacle that few wanted to miss. Aside from the affair being viewed as an excellent match up, most were also there to see the brash young Sir James put in his place. Pierre was easily everyone's favorite at court and they were pulling for him.

About the only two in the court who did not think Pierre would really take the gifted but unlovable knight was Sir James and Pierre, himself.

Pierre knew he had the skills to compete with the likes of the knight with the Crimson Shield but some said that the man from Bayard really did not fathom how good he himself, really was.

After helping fit the new knight into his borrowed

armor, Pierre Bellabre conducted a final inspection of all of Pierre's equipment. Then Bellabre took the green and white scarf from Pierre and secured it tightly around the knight's arm.

Then they heard the trumpet blast which signified that the Count had arrived at his box. Bellabre handed Pierre a lance and a shield.

"Good skill, my brother." offered Bellabre with a smile as he turned to lead the battle charger to Pierre's designated end of the list.

At the signal, the two combatants guided their mounts before d'Ligny's box and greeted the count.

"May God protect you both." said Count d'Ligny. "You may take your places and begin at your liking."

As Pierre turned his mount he could have sworn that he had detected a quick wink from the count.

The contestants rode a few yards together before they would part and go to their respective ends of the lists.

More than a little irritated at Pierre for having created this entire situation, Sir James snapped. "You foolish lad. The pity that you should begin knighthood in the dust from the stroke of my lance. I trust it will not embarrass you too much when you relate to the Lady Anne of this meeting and of who was the better man."

Pierre was feeling cocky, too. He resented being

called lad, especially by someone who was only four or five years older than himself. He knew that the knight with the Crimson Shield was only attempting to gain a psychological advantage. He decided to play a hand at the game himself.

"Of a truth, sir knight." said Pierre in a more jovial voice than he thought he would manage. "I trust you will well remember well what is about to happen when it is you who kneel before her."

Pierre's words had their desired effect on Sir James, who began to boil inside. "I was going to take it easy on you." snapped Sir James angrily. "But now I believe I will crush you soundly."

"Protect yourself, sir knight." Pierre replied with as much calmness as he could muster. "For the ground is hard this time of morning."

Bellabre shook his head as Pierre rode to his spot. "My brother, you are either very foolish or very brave for provoking him so before doing battle."

"Where is your faith, mon ami?" asked a chiding Pierre. "You worry too much. He is so angry by now that I doubt if he thinks too clearly. So, stand aside and signal that I am ready."

At the next blast of the horn the battle was on with the crowd screaming encouragement as the two riders prodded their mounts forward. Sir James was a well skilled knight and, despite his feud with

Pierre, had many backers. However, Pierre was much more admired by serf and nobleman, alike, as he had been, perhaps, the finest squire of any age. He drew a tremendous ovation as the galleries chanted "Picquez! Picquez!"

Bending low in his saddle, Pierre urged his battle charger to full speed.

When only a few yards separated the riders, the two lances dropped into the striking position. Seconds later, amid great cheering, both lances struck their targets. Both riders were sent reeling in their respective saddles but each managed to stay upright as they turned and readied for a second pass.

This time, at the very last possible instant, Pierre leaned in his saddle as he warded off Sir James' lance strike with his shield. In the same quick movement, Pierre aimed and struck. His own lance slammed against the Crimson Shield with stunning force.

There was complete silence, almost as if time itself stood still for that split second. Everyone strained to see the outcome of the clash.

To Pierre Bayard, there was never a doubt. He knew the instant that his lance struck the Crimson Shield that it was a good hit.

Sir James sat upright for a couple of seconds and then toppled, head over heels, backwards from his horse. As the knight with the Crimson Shield hit the ground with a resounding clanking thud, the gallery erupted in a wild ovation.

The thoughts in every head, at that moment, ran toward recognition that the new knight from Bayard was already of championship caliber.

Pierre turned his horse at the end of the lists so he could get a better look. He smiled proudly as he flipped up his visor and watched two squires and two other men rush to help the fallen knight regain his footing.

As Pierre Bayard was already one of the most skilled and accomplished soldiers in all of France, Sir James had no reason to be too humiliated, other than the damage to his personal psyche. It was not a disgrace to come in second to such as Pierre Bayard.

There was nothing left for Sir James to do except to swallow the bitter pill of defeat, put his pride aside and go to present himself before Anne de la Chavet to proclaim to her Pierre Bayard's deed.

CHAPTER TEN

It had been a particularly tiring day and the young knight from Bayard had just eased himself down to rest when the knock came upon his door.

"Enter!" he called out in a weary voice.

The young page opened the big wooden door to his quarters, holding a candle high to give the room more light.

"I beg your leave for disturbing your rest, Sir Bayard." said the boy. "A visitor awaits and said his business is too important to await the 'morrow."

"Methinks you should bid him enter, then." said Pierre as he sat up and lit the huge candle beside his bed.

When the sturdy figured entered the room, Pierre's muscles automatically flexed to full alertness as he recognized Sir James.

"Sir Knight." said Pierre. "Have you accomplished that which you set out to do?"

"As best as the circumstances would allow" answered the knight. "I am in the wrong in how I have been acting and I would like to put an end to our discord. I offer you my hand in friendship."

Pierre had to admit that he was a bit stunned by this sudden turn of character of the knight with the Crimson Shield. Yet, as he gazed into Sir Jame's eyes, Pierre felt the man was being sincere.

"I will accept your hand as it is offered." said Pierre after a long moment. "Now speak to me of that which I yearn to know."

Sir James looked at Pierre and his face grew strangely sullen. "It saddens my heart to bear you the tidings that I must, sir knight." said James quietly and with much hesitation.

"What is it? Tell me." prodded Pierre as impatience slipped into his voice.

"It concerns your lady love, sir knight." answered James. "She lies gravely ill. I could not really talk to her. She lays asleep. Sir Bayard, they informed me that she has been stricken with the plague."

THE PLAGUE.

Those two terrible words ripped through Pierre's heart like heated iron rods. Only a few years before had this dreaded killer struck and wiped out hundreds of thousands of people throughout Europe.

"Not Anne. Please, God, not Anne." whispered the stunned man from Bayard. Then, gathering himself, Pierre looked at Sir James. "Thank you for taking the trouble of coming to see me, sir knight."

"I am truly sorry. I will leave you in your hour of grief." said James quietly. Then he turned and was gone.

As soon as Pierre explained his situation to Count d'Ligny, he received permission to go to her on a twelve day leave from his duties.

After telling Bellabre and Captain d'Ars goodbye, Pierre put on his full body armor and took his pure white shield, as he had not yet chosen a coat of arms, and took to the road on Wingfoot. "Speed to you, great horse." called out Pierre. "We have a great mission to complete."

As if the mighty animal understood the urgency of his master's words, his strides seemed to lengthen as his pace grew faster.

Pierre could feel the cool breeze on his tear stained face as he rode. He scarcely took note of the blackness of the night as the radiant face of Anne seemed to be riding on the horizon just ahead of Wingfoot's flying feet. Her soft lips seemed to be calling to him; pleading with him to hurry.

"Dear God. Not Anne."

Pierre found himself whispering this phrase over and over as he rode. Deep into the night they went. Finally, even the great horse was spent. Not wanting to waste even a precious moment, Pierre dismounted and walked for what seemed like hours.

It was near daybreak when the blackness began to lose its battle with the upcoming morning light, when Pierre came to a small farm.

Seeing Pierre's golden spurs, the farmer was willing to do anything in his power to aid the knight of France. Spotting a huge wagon, Pierre emptied his purse into the farmer's hands and begged to rent the wagon for a day. When Pierre explained his mission and the reason why he could not lose sleeping time, the happy farmer gave Pierre and Wingfoot food and water.

Then, in possession of more money than he had probably seen in years, the farmer helped Pierre get Wingfoot into the big wagon and then he began to drive his two horses along the road. They pulled the huge wagon as the very weary horse and rider slept in the hay-covered wagon bed.

It was nearly noon when the farmer pulled the great wagon to a halt at a large stream. Pierre was instantly awakened. He let Wingfoot get his fill of water and eat some hay while the farmer gave him some bread and a drink of wine. Then the farmer did an unexpected thing as he gave half of Pierre's money back to him and babbled something about being in that kind of love, himself, before.

Throughout the day the mighty horse carried his rider ever closer to Anne.

As the evening shadows began to point their long, purple colored fingers toward earth, the darkening French sky became filled with storm clouds. An hour later, Pierre was blinking against the heavy downpour as he rode. Soon, the ground became muddy and even Wingfoot was forced to slow his pace considerably.

It was well after dark when the rain quit. Soaking wet, Pierre began to feel the wind as it chilled him to the bone. Wingfoot was spent and Pierre dismounted and then sloshed, doggedly, through the mud.

The Plague!

The words kept pounding at his weary brain. He fought the tears as he walked and kept whispering Anne's name. He dared not let himself dwell on the thought of death's cruel hand possibly snatching her away.

Pierre plodded for miles before remounting and urging the great horse onward again. They both were gathering their second winds as the surrounding countryside began to appear familiar. Shortly they passed the great estates of Bocsozell and Theys and rode on.

The bright moon intermittently peered out from behind the dark scattered clouds as the eastern sky took on the scant shape of morning.

On through the valley of Graisivaudan rode Pierre Bayard. Off to the north a few miles would be his home. But he could not go in that direction at the moment. He turned and took the rode leading toward Grenoble where he would find the home of Anne de la Chavet.

His bone weary body seemed to take on new life as he saw the walls surrounding his destination. When his voice commanded the guard to open the great gate a helmeted head peered over the rampart. One look at the knight and the head disappeared from sight. Then the gate swing open with a creaking sound. Pierre guided Wingfoot into the courtyard as a guard approached.

"An odd hour to be calling, Sir Knight." said the man.

"I am Pierre Bayard of the great house of d'Ligny." snapped a tired traveler. "I have come to see my beloved who lies ill within this place. Take me to the master of the house."

The man quickly led Pierre to the door and pulled on the rope that rang the bell inside. "Care well for my horse." said Pierre. "He has earned a good meal and a long rest."

As the door opened, the guard turned Pierre over to a sleepy eyed page who led him up a long flight of stairs and nodded him toward a doorway at the end of the hallway.

Pierre knocked softly on the door and waited until a voice bid him enter.

The knight removed his hood, letting his dark straight brown hair fall lightly about his ears. He entered and looked into the kindly face of the old man who sat beside the huge canopied bed.

The old man looked, curiously, at the tall young soldier with the dirt on his clothing.

"I am Pierre Bayard." said the knight in a quiet voice. "I… I have come to..."

A sign of acknowledgment crossed the old man's face as he heard the name. "I am Anne's father." he said softly.

"How is she?"

"She is not well." admitted the man. "She lies unconscious as you see her and has been so for six days and six nights now."

Pierre moved a couple of steps closer to the bed and looked at it for the first time. He grew heartsick at the sight of the figure laying there. She was barely visible to him by the light of the single huge candle. Pierre wanted to cry at seeing her. Her once peachy colored cheeks were now a pale and blanched grey white. Her once bright and sparkling eyes were now closed and motionless. The once smiling and happy face now lay in a state of non-expression.

Just for the briefest of moments, Pierre pretended that she was not sick, but only asleep.

His illusion, however, quickly ended as he heard the old man's voice again.

"It is a sad sight, indeed."

Anything he could say would have been futile so Pierre said nothing He could only stand and gaze, helplessly, down at the girl he loved. The old man was speaking again. "Her physicians have left her, saying that they can do no more. We have naught but to wait."

"Wait? Wait for what?" Pierre heard himself blurt out. "Wait for death's ugly hand to claim her?"

The old man started to respond. But, then, he looked sadly down at his daughter and then back, again, at the weary young knight. "Call me if there is any change." he said in a weary voice. "I must sleep now."

Pierre was barely aware of the man leaving as he stood for a long moment just looking at the figure in the bed. Then, slowly, he sank to his knees, all the weariness in him finally weighing him down. Pierre never took his eyes off the young girl as he fought the thoughts building up within him. No! His Anne wasn't going to die. She was so young. So pretty. So sweet of nature and kind of heart. Maybe this was not the sleeping death, he kept thinking. Maybe she was just asleep? Even as he tried hard to convince himself of those thought, Pierre knew deep in his heart that they were not true. Anne had a very slim chance, at best, to pull out of the grasp of this terrible thing which held her very life in its fingertips.

The room was hot and stuffy in these early morning hours. No wonder, he thought as he looked around for the first time. The place was boarded up like an old barn. Pierre forced himself to stand and move over to the opening in the wall. His steel hard grip and mighty arms forced the heavy wooden window open. The cool whisper of a breeze met his face with welcomed enthusiasm.

The slight sound of the rustling of sheets from the bed grabbed his full attention. Pierre whirled in anticipation and his tired eyes gazed intently at the still form in the bed.

Was she coming out of it?

Was her danger over with?

Was this the end of the terrible nightmare?

Three quick steps took him from the window to the bed, once more. For several moments that seemed like hours, he stood in silence gazing into the now pale face. The face that was once so full of life and had stirred excitement in the very core of his being each time it smiled at him.

Pierre, once more, knelt beside the bed. His usual steady hand trembled just a bit as it reached for the silken hair. Even in her present state of illness, Anne's hair was soft and silken to the touch. Her brow was moist with perspiration. Pierre looked around the room for something to mop the wetness away.

Then he remembered the green and white silken scarf that he always had resting near his own heart. He withdrew it from beneath his tunic and brushed it gently over her forehead. A couple of tears won the battle he was fighting within himself and rolled slowly down his suntanned cheeks.

Pierre then took Anne's hand in his and bowed his head in prayer. There in the darkness with only a small candle lighting the room, Pierre poured out his tearful prayers. For well over an hour Pierre prayed. Then the inevitable blanket of sleep spread over his weary body. The blackness crept into his eyes and blotted out all else.

The midnight hour came and went again. Even though he desperately needed sleep, something suddenly compelled him to open his eyes again and look at the girl. "Anne." he whispered. "My Anne! How I long to hold you and hear you laugh once more. Awaken and let me show you how I have fulfilled my... our dream."

Soon, though, the inevitable sleep surged up; again swarming over him. He forgot about the candle. He forgot about the soft breeze and the dull ache in his back from the kneeling position he had been in for so many hours now. The last thing that the weary eyes saw was the silken green scarf that he held in his left hand. Therefore, the young knight did not notice the bright moon as it peeped from behind the cloud and beamed gently into the opening in the room. The moon's rays spread over the two figures and the bed.

Her tired eyelids flickered weakly. Then, for the first time in countless days and nights, they fluttered open. Save for the dim moonlight which filtered through the open window, the room was dark. Slowly the gaze roamed as the tired eyes tried to focus on the familiar objects. Aware of the presence of someone else in the darkened room with her, the pale and very weak girl lifted her eyes in a sweeping search.

Then the gaze slowly settled on the suntanned dark face of her Page. But, something was different about him. Wait! He was no longer dressed as a Page. He was all grown up. Yes! He had become a squire, the tired mind reasoned. She remembered the letter now. How proud she had been of him.

But, no! Wait!

Pierre was not dressed as a squire. She was confused now as she tried hard to concentrate on the figure close to her. Then she realized that his clothing was that of a full-fledged knight.

A weak, yet unmistakable, smile slowly crossed her pale lips as she then saw the familiar green and white scarf that he held in his hand. Calling on every ounce of strength she could muster, Anne slowly reached a trembling hand for the pretty scarf as a name softly escaped her mouth.

"Pierre."

Then the eyelids flickered, once more, and the eyes starred ahead into nothingness.

Pierre did not know what suddenly jarred him from his sleep. It sounded like a voice calling his name softly, he thought. His sleepy eyes tried to focus in the dark as the moon had, again, drawn the blanketed cloud over its face.

With fumbling fingers, Pierre hurriedly lit another candle, trying to cast more light into the room.

Pierre's first quick glance around the room was sufficient to let him know that they were still alone.

But, if this were true, then who spoke his name?

Had it been a dream? Was it something that he had wanted so desperately that he had conjured it up in his sleep?

The young knight took another look at the girl in the bed.

Her smile went deeply into his heart, as he really looked at her face. It was as if she were trying to tell him something; trying to give him some kind of a hidden message.

Then, suddenly, the horrible truth struck at him like a lightning bolt. Something else had changed in the look of Anne de la Chavet.

Pierre's blood froze when he saw her eyes.

They were OPEN!

They were open but she wasn't blinking.

Great swelling tears erupted inside of him as he realized that his Anne was DEAD.

Pierre's trained eye also fell on the green and white silken scarf that was tightly locked in the fingers of the dead girl's hand and was now crushed tightly to her bosom.

Pierre Bayard burst into uncontrollable tears and he collapsed on the bed; his head coming to rest atop the green and white scarf.

CHAPTER ELEVEN

After Anne had been properly buried, it was with a sad and heavy heart that young Pierre Bayard returned to his duties in Lyons.

Now, as a young lance, Pierre Terrail Bayard lived a much different life. As a lance, too, he now drew an annual pay of 300 livres and free maintenance of three horses. He also was entitled to two personal archers, a valet d'arms, a page and a man servant. As a free roaming soldier, Pierre now belonged to the body of fighting men who were directly dependent on and paid for by the king. They formed the bulkhead of the first standing army that Europe had ever seen. This army had but the single purpose of protecting the king's land and keeping peace.

The size of each elite company in this army was dictated by the size of the particular garrison they manned or by the task they undertook. These companies ranged from thirty to one hundred men-at-arms. Each company was commanded by a battle hardened and very capable Captain, who served his noble prince, who governed the companies in nominal fashion.

It was another grand and proud day in the court of d'Ligny when, in late September of 1491, Pierre Bellabre, Gaspard De Coligny and Lucquin Le Groing all obtained their knighthood.

Coligny left Lyons and was assigned to some special duty with a powerful member of his family at the King's own court. This left the other three young friends still together at the court of d'Ligny. Soon, the trio of new knights became the talk of all Lyons as they rode everywhere together while performing their duties.

On February 7, 1492, they all accompanied Paul of Luxembourg to St. Denis to witness the crowning of the new Queen of France, Anne of Brittany. There, King Charles requested Pierre to give a special performance of his riding skills for Her Majesty, who thoroughly loved what she saw.

The new queen seemed nice enough, despite her short, thin body and readily noticeable lameness. Both of the Pierres thought that she had a much too stubby of a nose which seemed to hang over a much too spacious mouth and her long face only added a sickly look to her already pale complexion.

Yet, as Bellabre pointed out, she fit rather nicely into the royal court with the young king, who was already suffering from an abnormally large humpback, hesitant speech and a most unbecoming spasmodic twitching of his hands.

In spite of all this, though, Charles was quite proud of this marriage as it was most difficult to arrange. He had pulled it off in brilliant fashion. It was an important union for both France and Brittany.

King Charles was doing something else, too, in setting about to make a concentrated effort to solidify his standings with various members of his realm. In order to clear the way for the espousal of Anne, Charles had to send Marguerite of Austria back to her father along with an adequate compensation for 'dented' pride.

The king also, personally, went to the prison and released his cousin, Louis of Orleans, whom his cousin, Anne had imprisoned there a few years before. All in all, for a young king of only twenty one years of age, Charles' accomplishments were turning out to be of no small consequence.

The young king, despite his handicaps, was reputed to be more than adequate at tennis, tilting and hunting. For better or for worse, the monarch spent a great deal of his time at these passions. He also tried to be a good husband by showing his new Queen off to various parts of his realm. Lyons was one of his favorite places to visit and he often journeyed there from Amboise, Chateau Briant and Paris.

The last day of June of that year, 1492, proved to be a great turning point in the lives of the two Pierres and Luc Le Groing

Pierre received word in mid-morning that he had been requested to be in audience with the Count and his most special guest, King Charles, himself.

"What did you do, mon ami?" chided Bellabre. "Shoot a royal cow?"

"Come with me. Let us find out together." offered Pierre jokingly.

"No, thank you." replied Bellabre. "I can manage to get in enough trouble all by myself. Besides, I do not like the way of the courtier. You do not either. Why are you going?"

"Because my lord, d'Ligny had requested it." reminded Pierre.

"Ah, then it is best you do not terry." called out Bellabre as he walked away.

"Sir Bayard." greeted the warm voice of Count d'Ligny when Pierre was shown into the room. Pierre was surprised to see King Charles and dropped to one knee to pay him proper respect. When the king bid him arise, Count d'Ligny began to talk.

"You have surely accomplished much in your short career. But, methinks it is naught when compared to what you can accomplish in your lifetime as a soldier. Even though you are new in the ranks, you are proven to be one of the most accomplished soldiers under my command.

However, His Majesty holds you in much esteem and wishes for your military career to begin in earnest. Perhaps that would not happen if you stayed here putting down small outbursts of rabble in the hills. Therefore, His Majesty has desired that you leave Lyons."

Leave Lyons, thought Pierre? Was he really ready to do that, he wondered. Was he really ready to go into the field where real war could exist at a day's notice? The count's voice jarred him back to the moment.

"You, therefore, will be assigned to the garrison of Picardy near Aires. It could become a very important place should the trouble that has been brewing actually break out along the border." the count was explaining. You will join the garrison, which will be captained by my lord, d'Ars. The young gentlemen, Bellabre and Le Groing will be joining you in a few days."

King Charles smiled and spoke for the first time.

"Young sir, you are quite an extraordinary... fellow." began the king as his hesitant speech pattern began to show, even though he tried hard to control it. "I... I would have it no oth..other way than to present you… with a gi..ft of three hundred crowns.. And a strong..bat...battle steed."

"Your Majesty is much too kind." said Pierre. "But this humble soldier thanks you."

"Little enough..prize for you..Picquez." said the king with a genuine smile. Charles nodded to d'Ligny; the signal that he was finished speaking. Paul of Luxembourg took over.

"I, too, would not think of letting you depart without offering you a gift." said the count. "To spare your own fine steed, I wish to add, yet, another. I also will have sent to your lodging two suits of velvet."

D'Ligny, even though only a few years older than Pierre, spoke to him as a father, warning him of some of the things to expect at his first garrison. "Hold your honor above all things." suggested d'Ligny, knowing that the man from Bayard did that anyway.

Early the following morning, the excited young lance set out with his horses and servants for his first garrison. Bellabre accompanied his friend for a league or so until they pulled up at the main fork in the road.

With a wide smile, Bellabre pulled off his glove and extended his hand.

"Pierre, my fellow lance and dearest friend." said Bellabre. "I bid you safe and speedy journey. I will be unhappy until the day I can join you in garrison." Bellabre stopped and laughed his good natured laugh. "But then, that will afford you ample time to tell all the ladies of my coming."

"Bellabre, my brother." said Pierre with a smile. "You are beyond my comprehension. But I will be looking forward to seeing your face in Picardy."

Bellabre replaced his glove and wheeled his horse. "Well, if you keep me here talking, you will never leave." He waved over his shoulder as he rode away.

Pierre smiled to himself then waved his little group forward, thus, setting off for his very first garrison. Of course, neither of them knew that Pierre Bayard would, eventually become the most honored knight in any army on earth.

THE END

Other exciting adventures by Duane R. Ethington that you will want to add to your collection:

Limpy

His Brother's Keeper

The Last Desperate Mile

Moranbong

Nosslived

Young Bayard

Made in the USA
Middletown, DE
05 November 2022